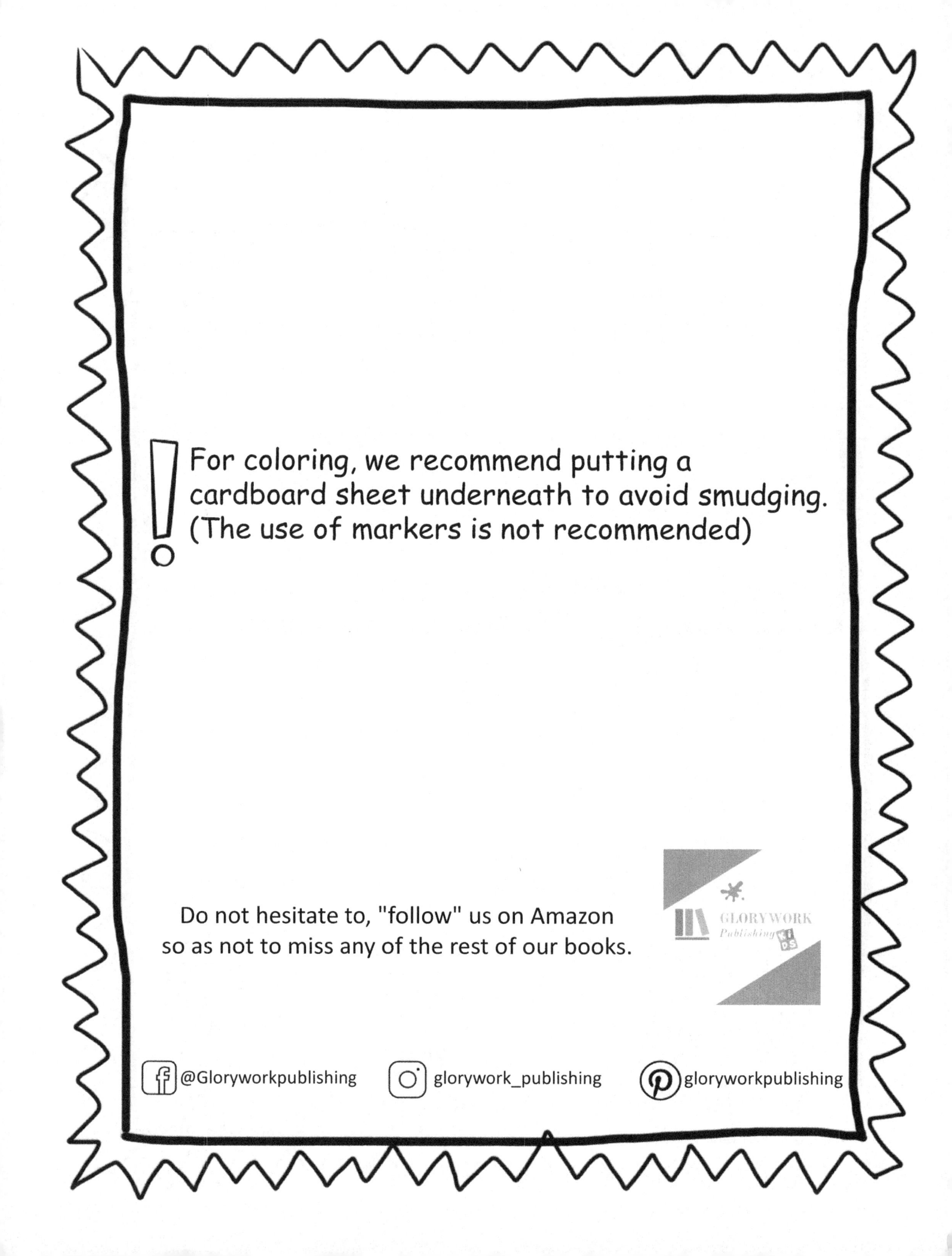

⚠ For coloring, we recommend putting a cardboard sheet underneath to avoid smudging. (The use of markers is not recommended)

Do not hesitate to, "follow" us on Amazon so as not to miss any of the rest of our books.

GLORYWORK Publishing KIDS

f @Gloryworkpublishing   ⓘ glorywork_publishing   ⓟ gloryworkpublishing

Singapore Math Kindergarten Workbook

Learn basics of math are foundational skills that form the basis of your child's academic future.

These 101 pages resources helps your child to master math skills such as number recognition, addition and subtraction, measuring, time and more.

This book is based on a mastery-focused approach called the Singapore math method, which is achieved through intentional sequencing of concepts. Some of the key features of the approach include the CPA (Concrete, Pictorial, Abstract).

This book is perfect for **kids kindergarten (ages 4-6)** but also is perfect for **1st Grade (ages 6-7)** who want to learn and improve their math skills.

Dear Parents,

We sincerely hope you find our kindergarten math book to be helpful and fun.

You can write to me at gloryworkpublishing@gmail.com with the subject ' kindergarten math' to get free printable practice sheets.

We would be very grateful if you could leave us a kind review and comment on Amazon, that will encourage me to make more books like this.

Thank you

Luciana

All rights reserved © Glorywork Publishing

This workbook is divided into following parts:

Part 1 : Exploring Numbers to 10
Number recognition, practice tracing numbers and learning to count objects

Part 2 : Comparing and Ordering Numbers
Less than, Greater than, Equal to
Ordering numbers from the smallest to biggest

Part 3 : Exploring Shapes and Patterns
Sorting, classifying and shapes names

Part 4 : Exploring Numbers to 20
Practice tracing numbers and learning to count objects

Part 5 : Addition and Subtraction
Number Bonds
Learning simple addition and subtraction

Part 6 : Numbers to 100
Exploring numbers to 100 and practice tracing

Part 7 : Measuring (size, length and weight)
Describe measurable attributes of objects
(Small/Big, Short/Tall, Heavy/Light)

Part 8 : Time
Telling time using the clocks (Analog Clock/Digital Clock)

0　1　2　3　4　5　6　7　8　9　10

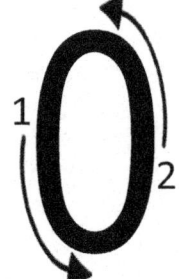 zero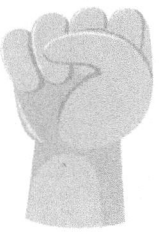

✏️ Trace the number 0

0   1   2   3   4   5   6   7   8   9   10

✏️ Color the number 0

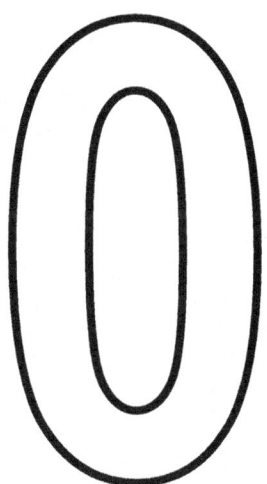

✏️ Trace the word **zero**

zero   zero   zero   zero

zero   zero   zero   zero

0 (nothing)

0   1    2   3   4   5   6   7   8   9   10

1 one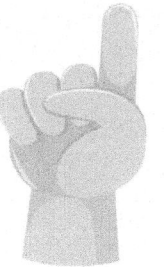

✏️ Trace the number **1**

0   1  2   3   4   5   6   7   8   9   10

✏️ Color the number **1**

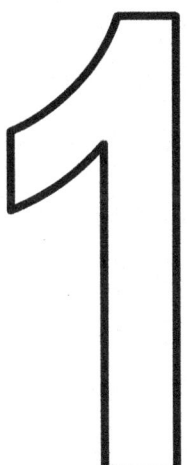

✏️ Trace the word **one**

one   *one*   *one*   *one*

one   *one*   *one*   *one*

1 Car

0   1   2   3   4   5   6   7   8   9   10

 two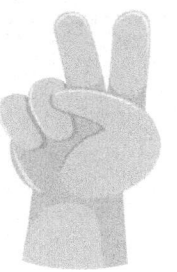

✏️ Trace the number **2**

0   1   2    3   4   5   6   7   8   9   10

✏️ Color the number **2**

2

✏️ Trace the word **two**

two   *two*   *two*   *two*

two   *two*   *two*   *two*

2 Cows

0 1 2 3 4 5 6 7 8 9 10

three

Trace the number **3**

3 3 3 3 3 3 3 3 3

3 3 3 3 3 3 3 3 3

3 3 3 3 3 3 3 3 3

3 3 3 3 3 3 3 3 3

3 3 3 3 3 3 3 3 3

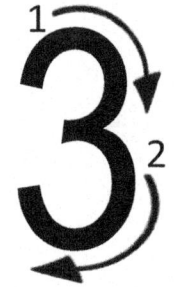

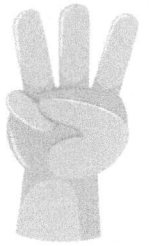

0 1 2 3  4 5 6 7 8 9 10

✏️ Color the number **3**

✏️ Trace the word **three**

three three three three

three three three three

3 Umbrellas

0   1   2   3   4   5   6   7   8   9   10

 four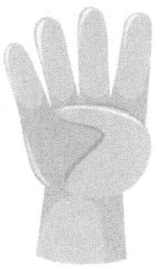

✏️ Trace the number **4**

0　　1　　2　　3　　4　　5　　6　　7　　8　　9　　10

✏️ Color the number **4**

✏️ Trace the word **four**

four　four　four　four

four　four　four　four

4 Tortoises

0   1   2   3   4   5   6   7   8   9   10

 five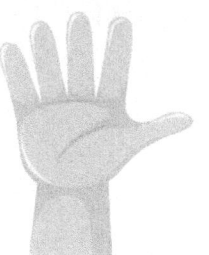

✏️ Trace the number **5**

0   1   2   3   4   5   6   7   8   9   10

✏️ Color the number **5**

5

✏️ Trace the word **five**

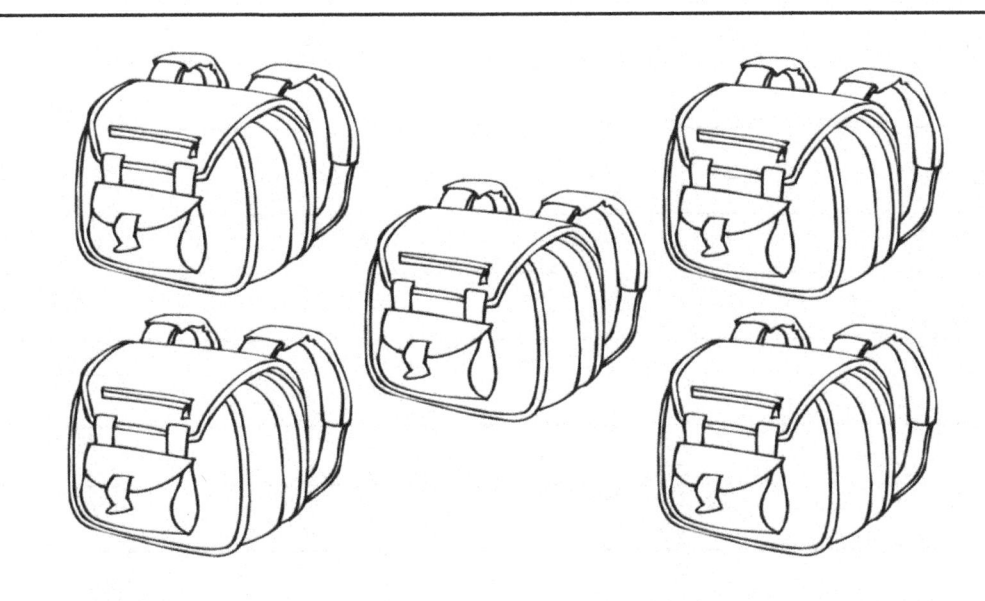

five   five   five   five
five   five   five   five

5 School bags

0 1 2 3 4 5 6 7 8 9 10

six

Trace the number **6**

6 6 6 6 6 6 6 6 6

6 6 6 6 6 6 6 6 6

6 6 6 6 6 6 6 6 6

6 6 6 6 6 6 6 6 6

6 6 6 6 6 6 6 6 6

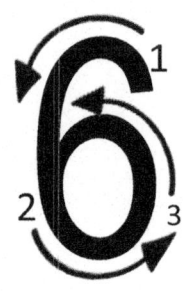

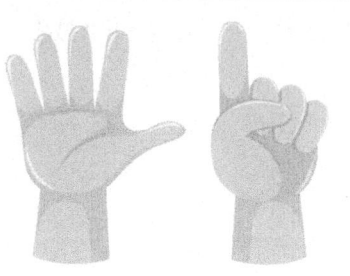

0   1   2   3   4   5   6   7   8   9   10

✏️ Color the number **6**

✏️ Trace the word **six**

six   six   six   six

six   six   six   six

6 Gifts

0   1   2   3   4   5   6   7   8   9   10

 seven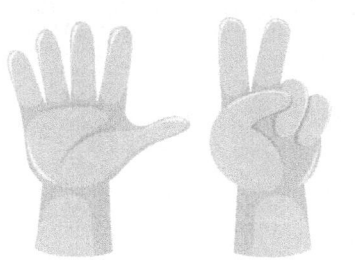

✏️ Trace the number 7

7 7 7 7 7 7 7 7 7

7 7 7 7 7 7 7 7 7

7 7 7 7 7 7 7 7 7

7 7 7 7 7 7 7 7 7

7 7 7 7 7 7 7 7 7

0   1   2   3   4   5   6   7  8   9   10

✏️ Color the number **7**

✏️ Trace the word **seven**

seven   *seven*   *seven*

seven   *seven*   *seven*

7 Flowers

0 1 2 3 4 5 6 7 8 9 10

eight

Trace the number **8**

8 8 8 8 8 8 8 8 8

8 8 8 8 8 8 8 8 8

8 8 8 8 8 8 8 8 8

8 8 8 8 8 8 8 8 8

8 8 8 8 8 8 8 8 8

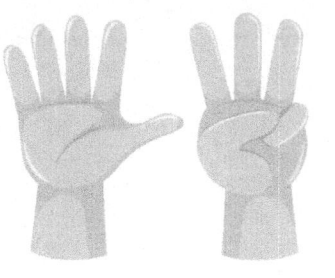

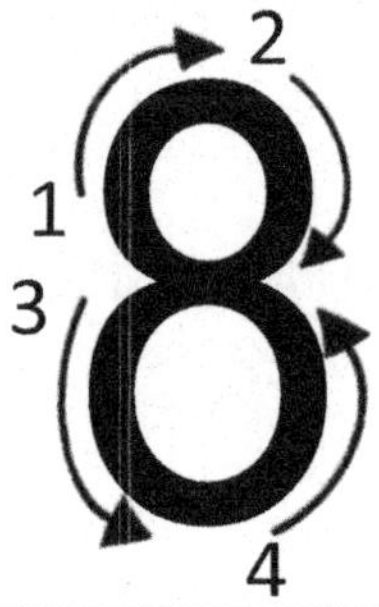

0   1   2   3   4   5   6   7   8  9   10

✏️ Color the number **8**

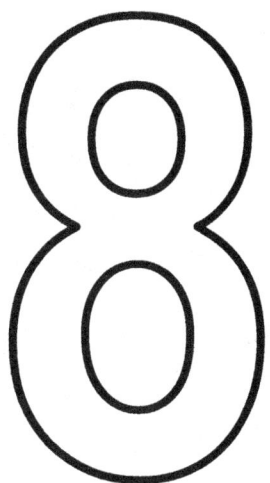

✏️ Trace the word **eight**

eight   eight   eight   eight

eight   eight   eight   eight

8 Pens

0   1   2   3   4   5   6   7   8   9   10

   nine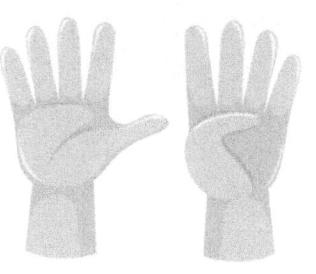

✏️ Trace the number **9**

0 1 2 3 4 5 6 7 8 9  10

✏️ Color the number **9**

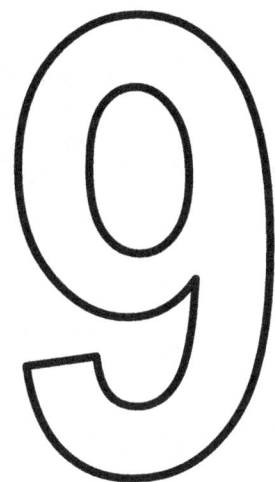

✏️ Trace the word **nine**

nine   nine   nine   nine

nine   nine   nine   nine

9 Butterflies

0 1 2 3 4 5 6 7 8 9 10

10 ten

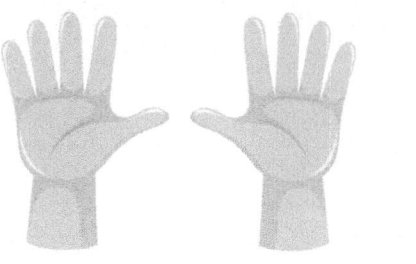

✏️ Trace the number **10**

10 *10* *10* *10* *10* *10*

10 *10* *10* *10* *10* *10*

10 *10* *10* *10* *10* *10*

10 *10* *10* *10* *10* *10*

10 *10* *10* *10* *10* *10*

0   1   2   3   4   5   6   7   8   9   **10**

✏️ **Color the number 10**

10

✏️ **Trace the word ten**

ten   *ten*   *ten*   *ten*

ten   *ten*   *ten*   *ten*

10 Koala bears

 Discover

1	
2	
3	
4	
5	
6	
7	
8	
9	
10	

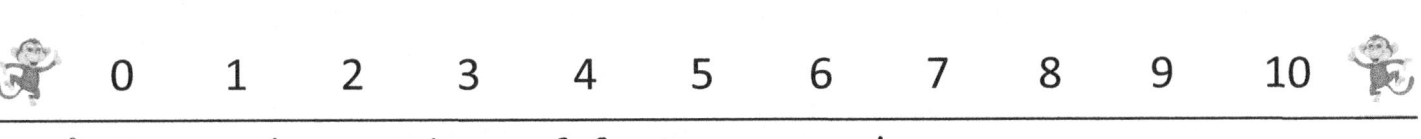

✏️ Count the number of fruits on each row.
Draw a line to match it with the correct number.

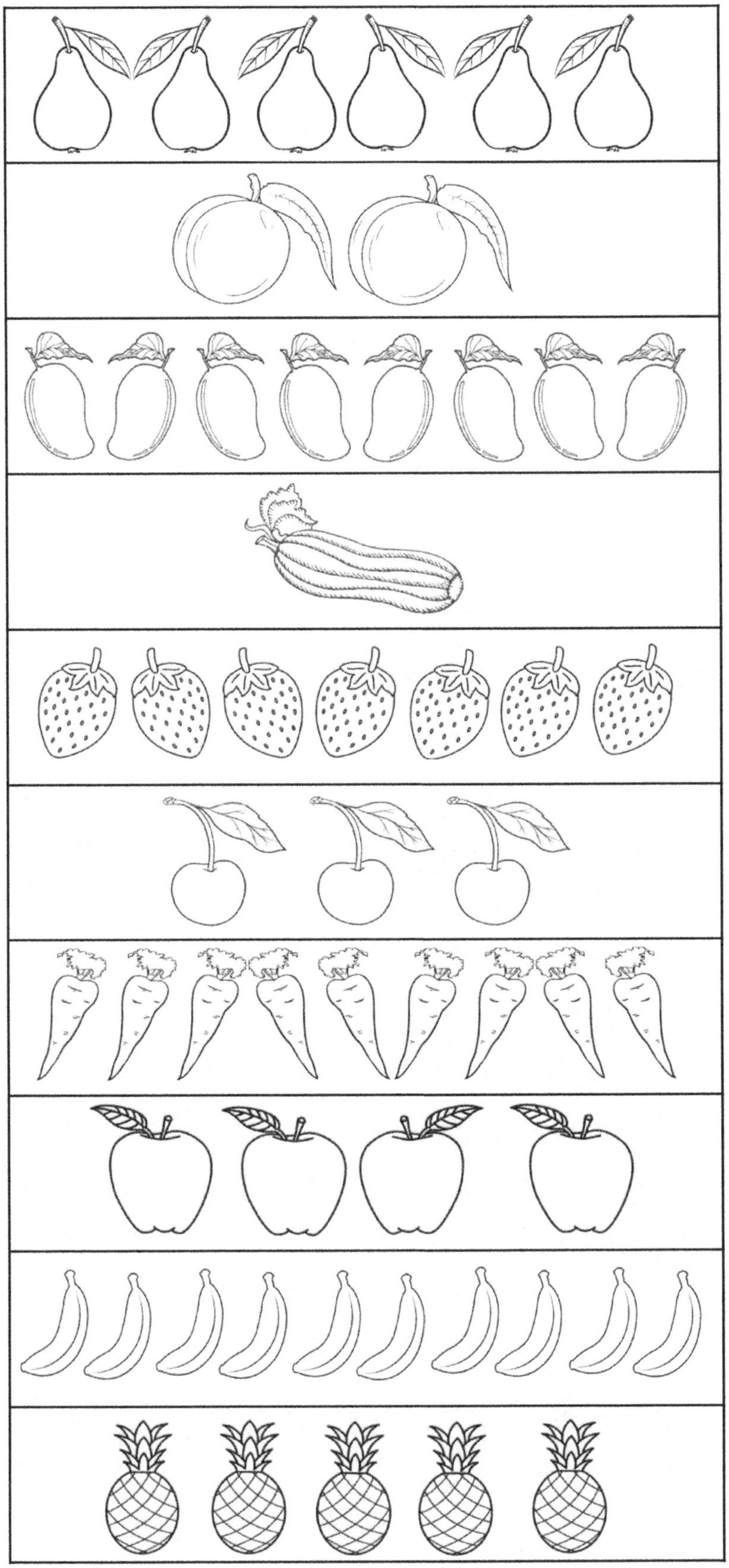

pears	• •	8
peaches	• •	3
mangoes	• •	7
eggplant	• •	6
strawberries	• •	10
cherries	• •	1
carrots	• •	5
apples	• •	2
bananas	• •	9
pineapples	• •	4

0 1 2 3 4 5 6 7 8 9 10

✏️ Colour the correct number of objects.

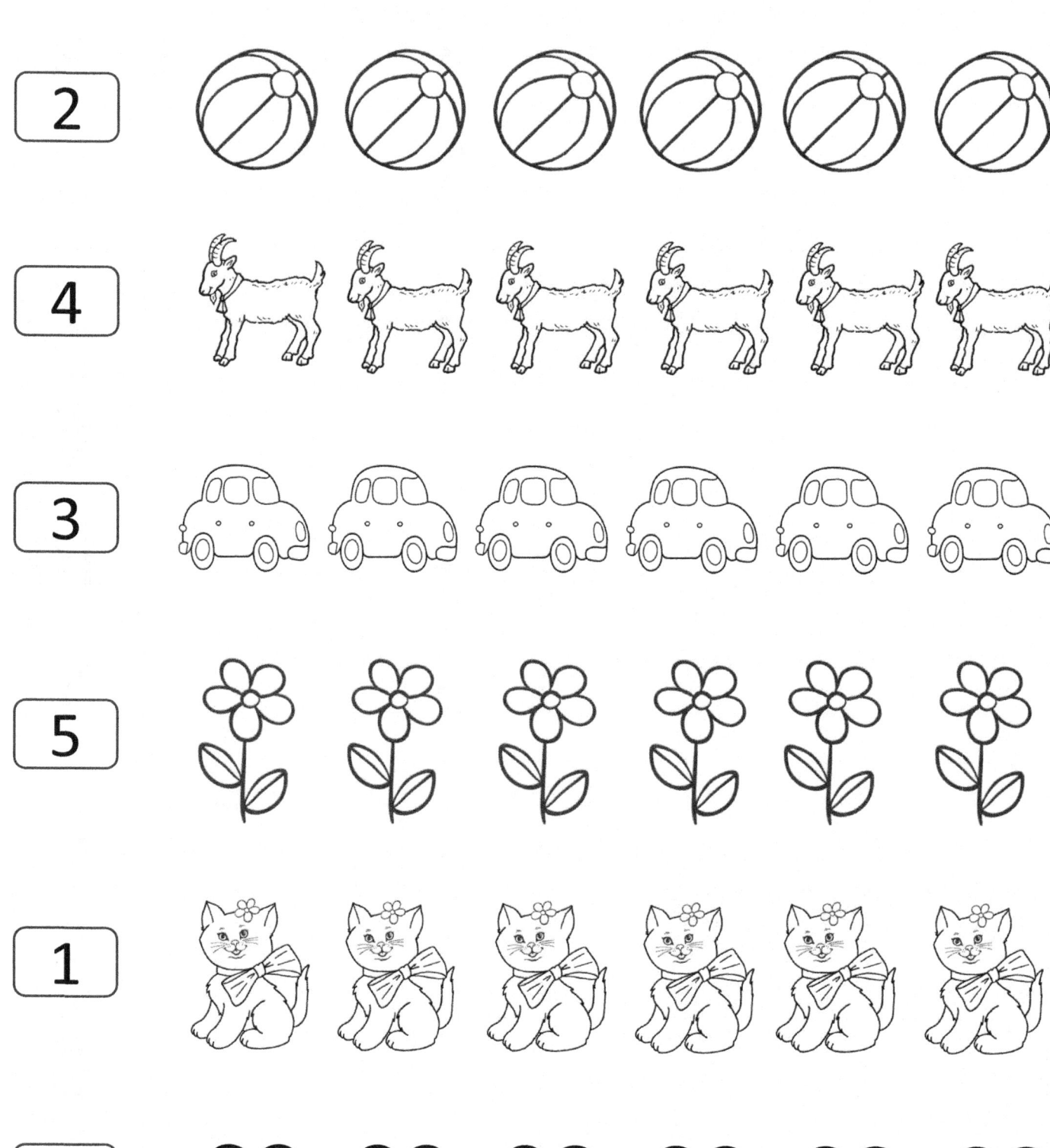

 0 1 2 3 4 5 6 7 8 9 10

 Count and write the number of objects.

3

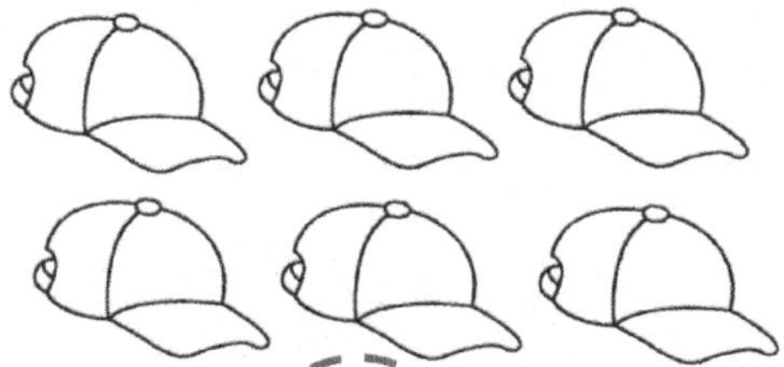

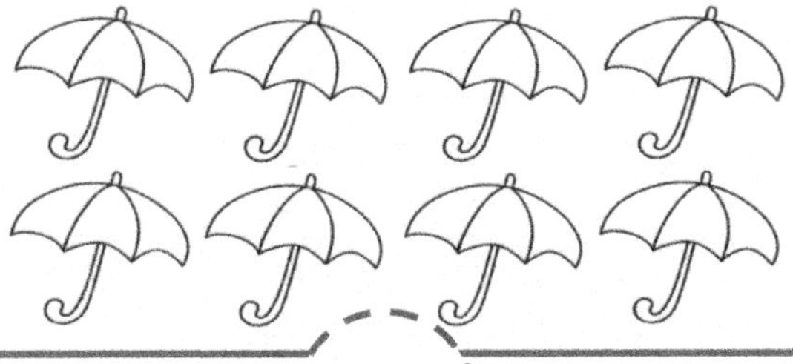

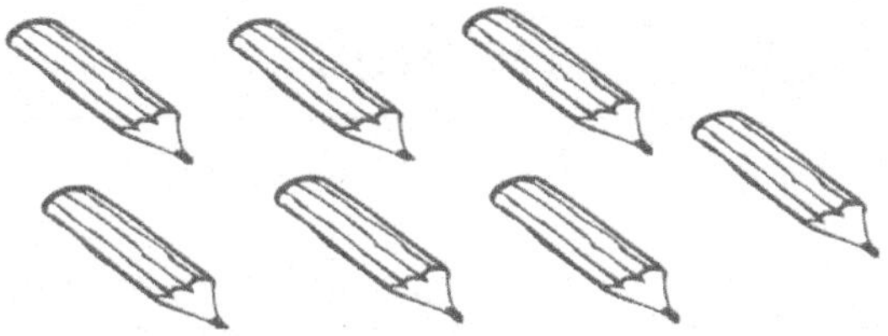

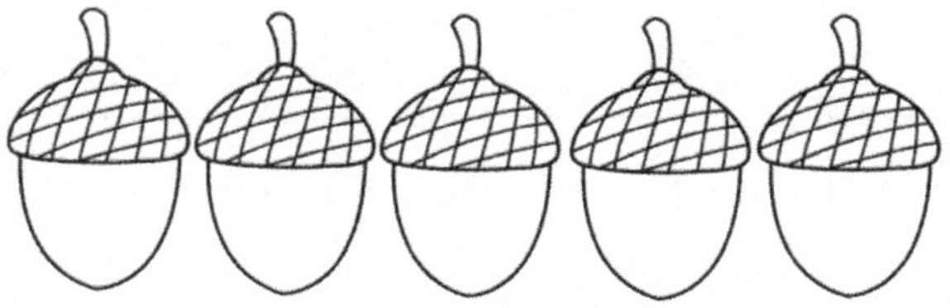

Ordering and Comparing Numbers

💡 **Discover** (Less than, Greater than, Equal to)

7 Candies   **is greater than**   3 Candies

7 > 3

3 Candies   **is less than**   7 Candies

3 < 7

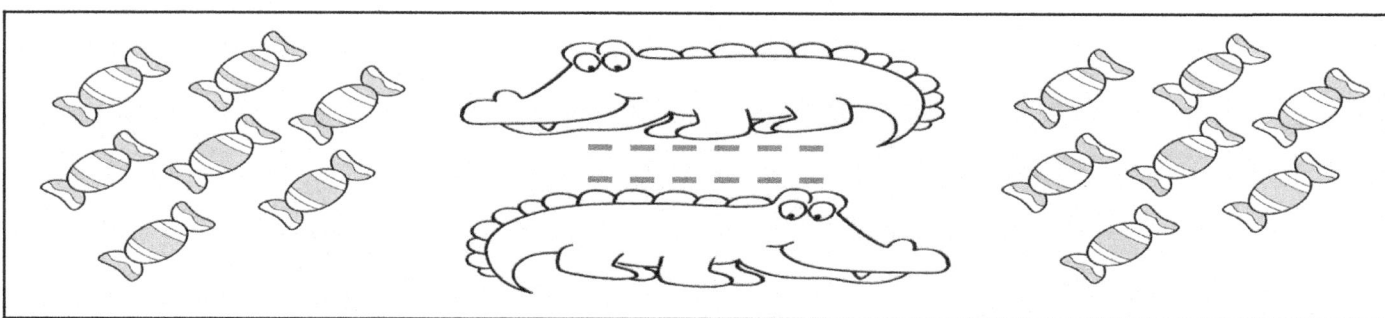

7 Candies   **is equal to**   7 Candies

7 = 7

>	<	=
is greater than	is less than	is equal to

Ordering and Comparing Numbers

✏️ Count the objects and circle the correct symbol: > or < or =

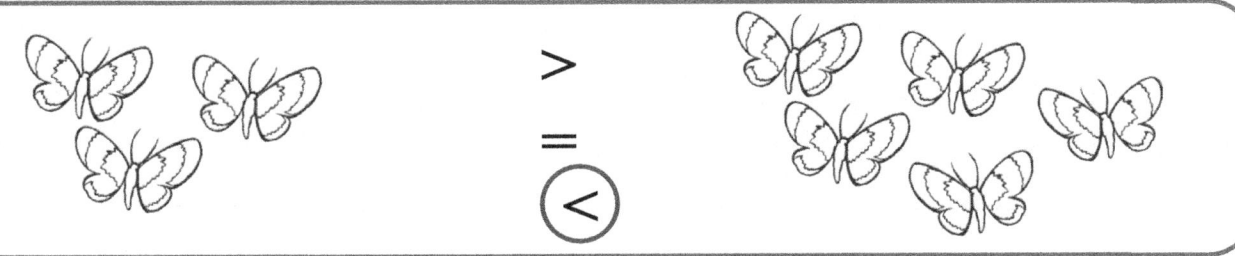

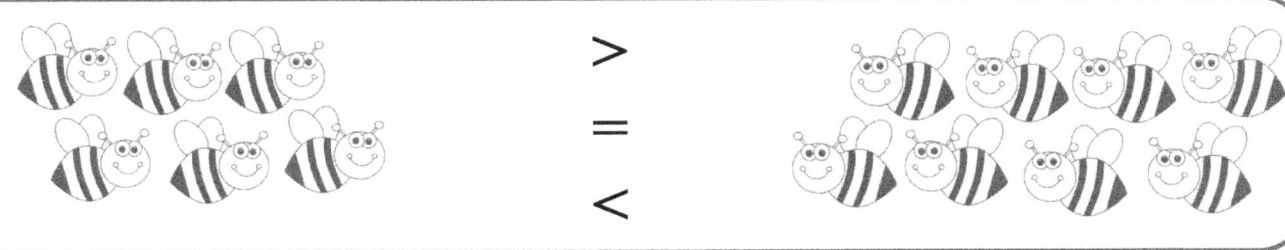

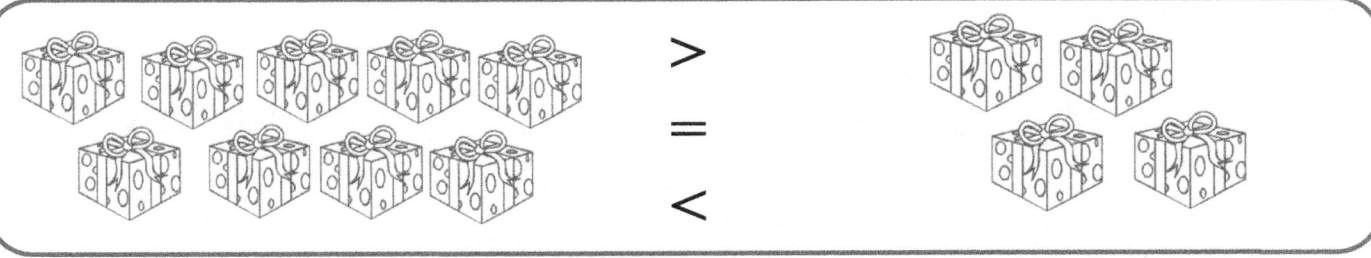

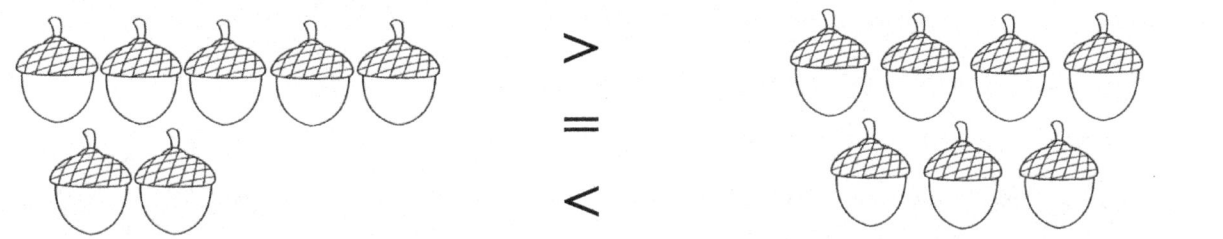

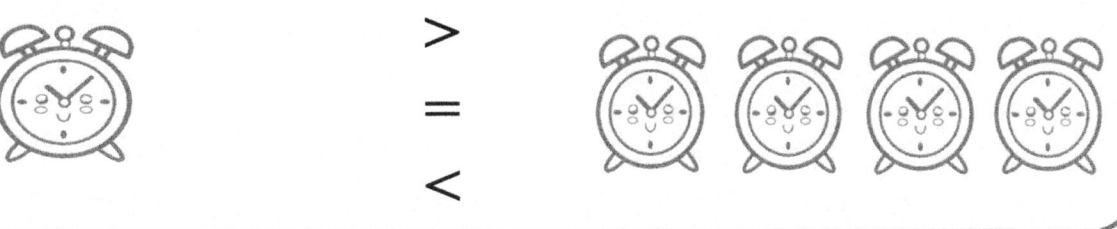

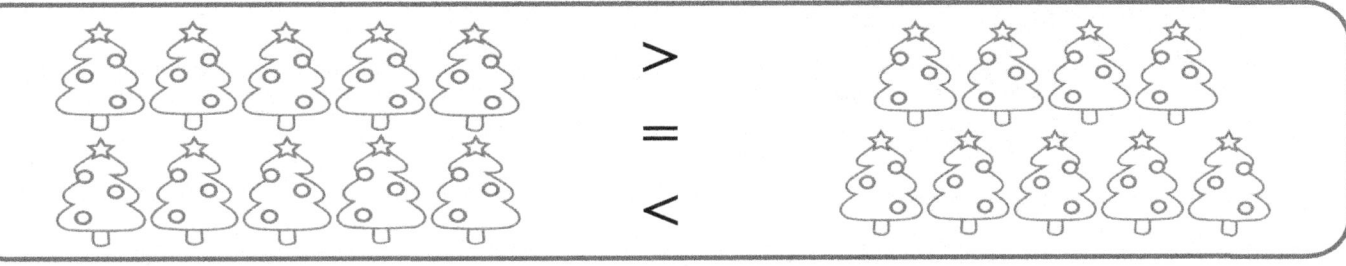

Ordering and Comparing Numbers

 Count the pictures and write the correct symbol >, <, = in each circle

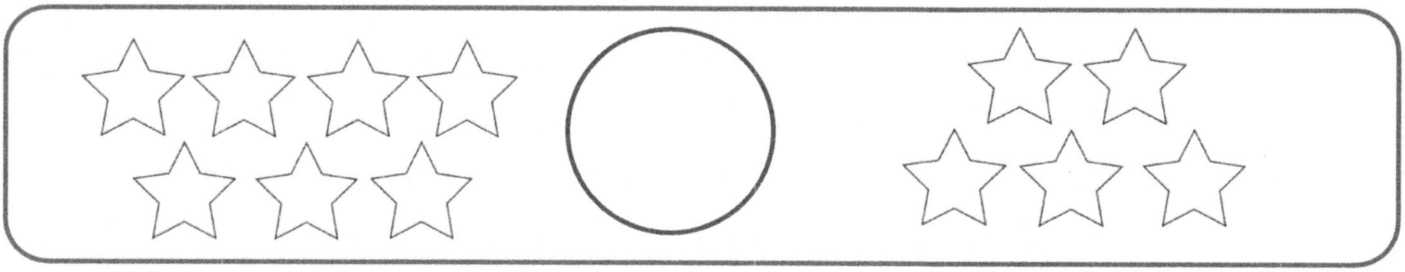

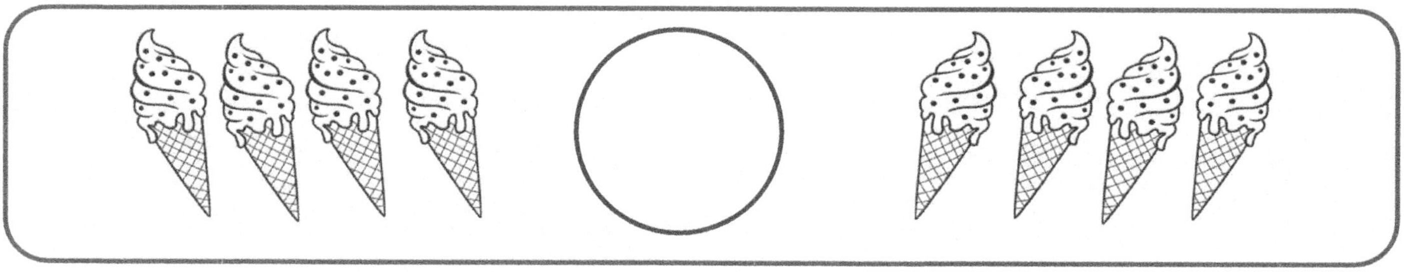

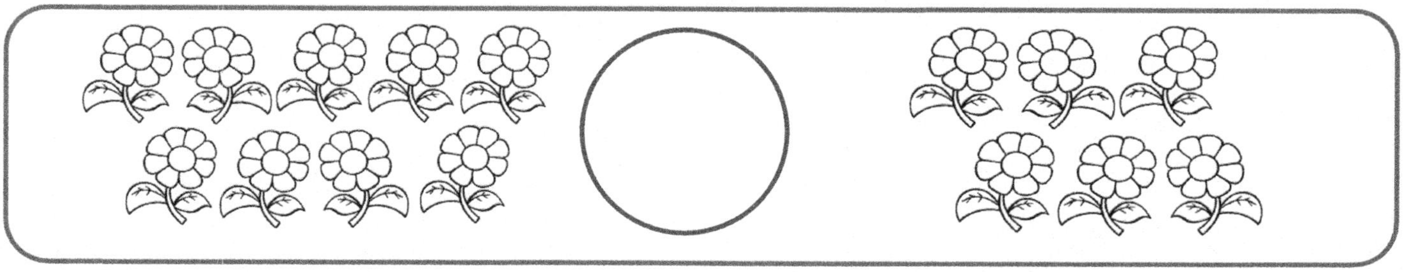

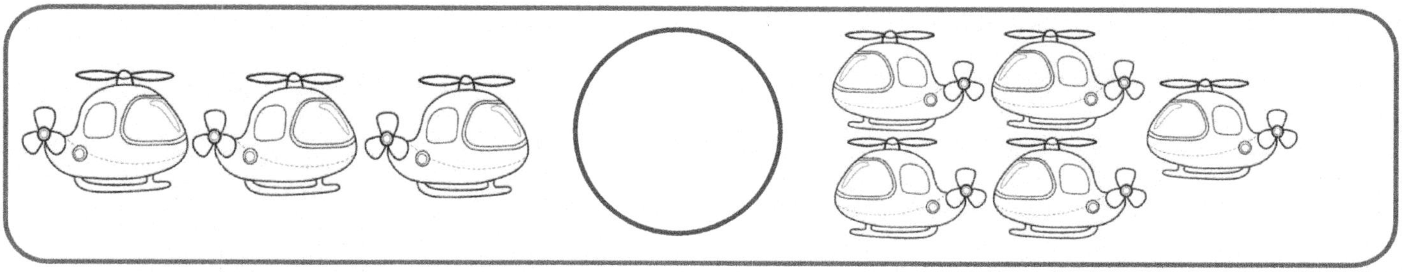

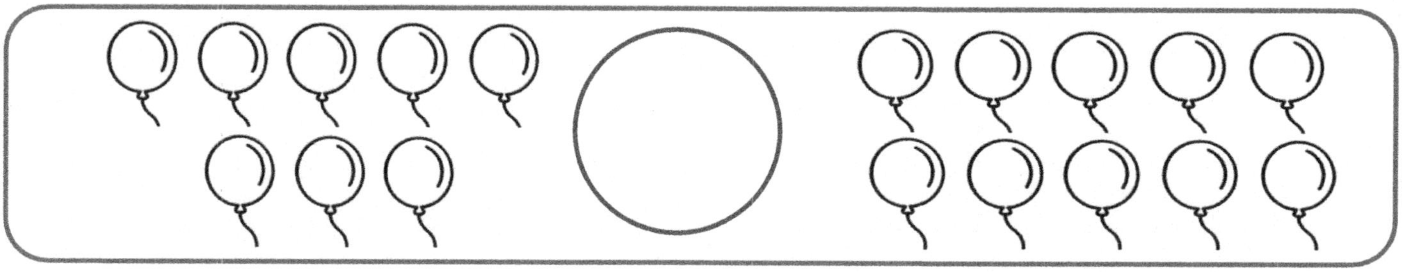

Ordering and Comparing Numbers

 Circle the number which **is greater** in each group

2 - 0 - 1

5 - 2 - 9

3 - 6 - 7

8 - 10 - 6

 Circle the number which **is less** in each group

10 - 3 - 4

2 - 8 - 5

3 - 1 - 6

9 - 10 - 8

Ordering and Comparing Numbers

 Ordering Numbers : From the smallest to biggest

5 - 9 - 7 - 2

What is the smallest number ? is 2

5 - 9 - 7 - 2

What is the next smallest number ? is 5

5 - 9 - 7 - 2

What is the next smallest number ? is 7

5 - 9 - 7 - 2

What is the last number ? is 9

5 - 9 - 7 - 2

Ordering and Comparing Numbers

❑ Compare the numbers using the correct symbol (>, <, =).

❑ Order the numbers.

2 ◯< 5

6 ◯ 1

4 ◯ 7

9 ◯ 9

5 ◯ 8

7 ◯ 9

10 ◯ 3

6, 3, 8, 1
1, 3, 6, 8

4, 2, 7, 6
___, ___, ___, ___

3, 0, 2, 8
___, ___, ___, ___

5, 10, 8, 1
___, ___, ___, ___

7, 9, 5, 4
___, ___, ___, ___

Shapes and Patterns

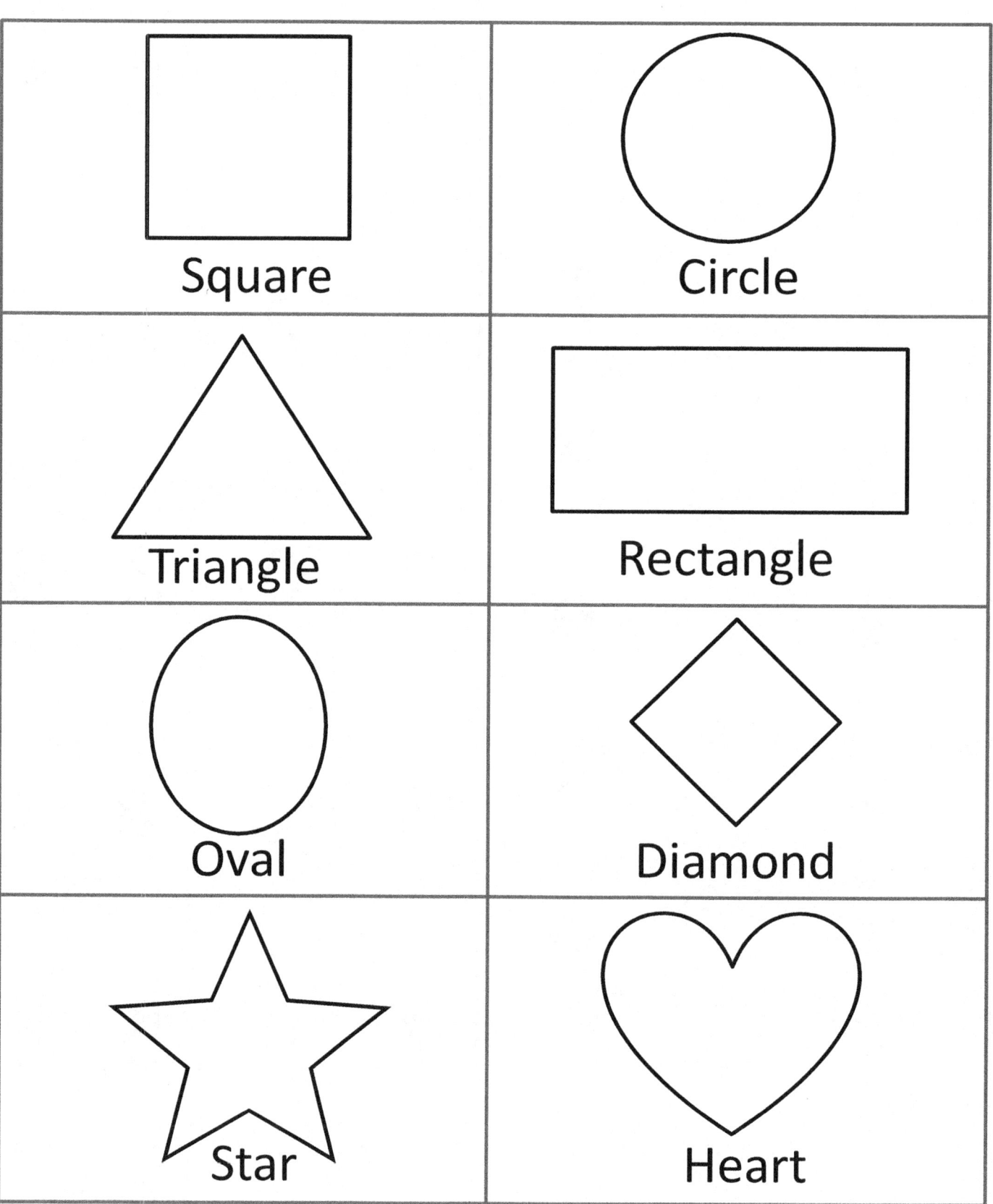

Discover

Square	Circle
Triangle	Rectangle
Oval	Diamond
Star	Heart

Shapes and Patterns

✏️ What shape comes next?

1. ○ ♡ ○ ♡ ○ ♡ _____

2. □ □ △ □ □ △ _____

3. □ ☆ △ △ □ ☆ _____

4. ○ △ ▽ △ ○ △ _____

5. ▯ ▭ ▯ ▭ ▯ ▭ _____

6. ♡ □ ▭ △ ☆ ♡ _____

Shapes and Patterns

✎ Colour the squares green, circles orange, triangles purple, rectangles pink, stars yellow, hearts red, ovals beige and diamonds blue.

How many of each shape did you find?

Shape	Square	Circle	Triangle	Rectangle	Star	Heart	Oval	Diamond
How many?								

Shapes and Patterns

✏️ Trace and color the shapes.

11 12 13 14 15 16 17 18 19 20

11 eleven

✏️ Trace the number **11**

11 11 11 11 11 11

11 11 11 11 11 11

11 11 11 11 11 11

11 11 11 11 11 11

11 11 11 11 11 11

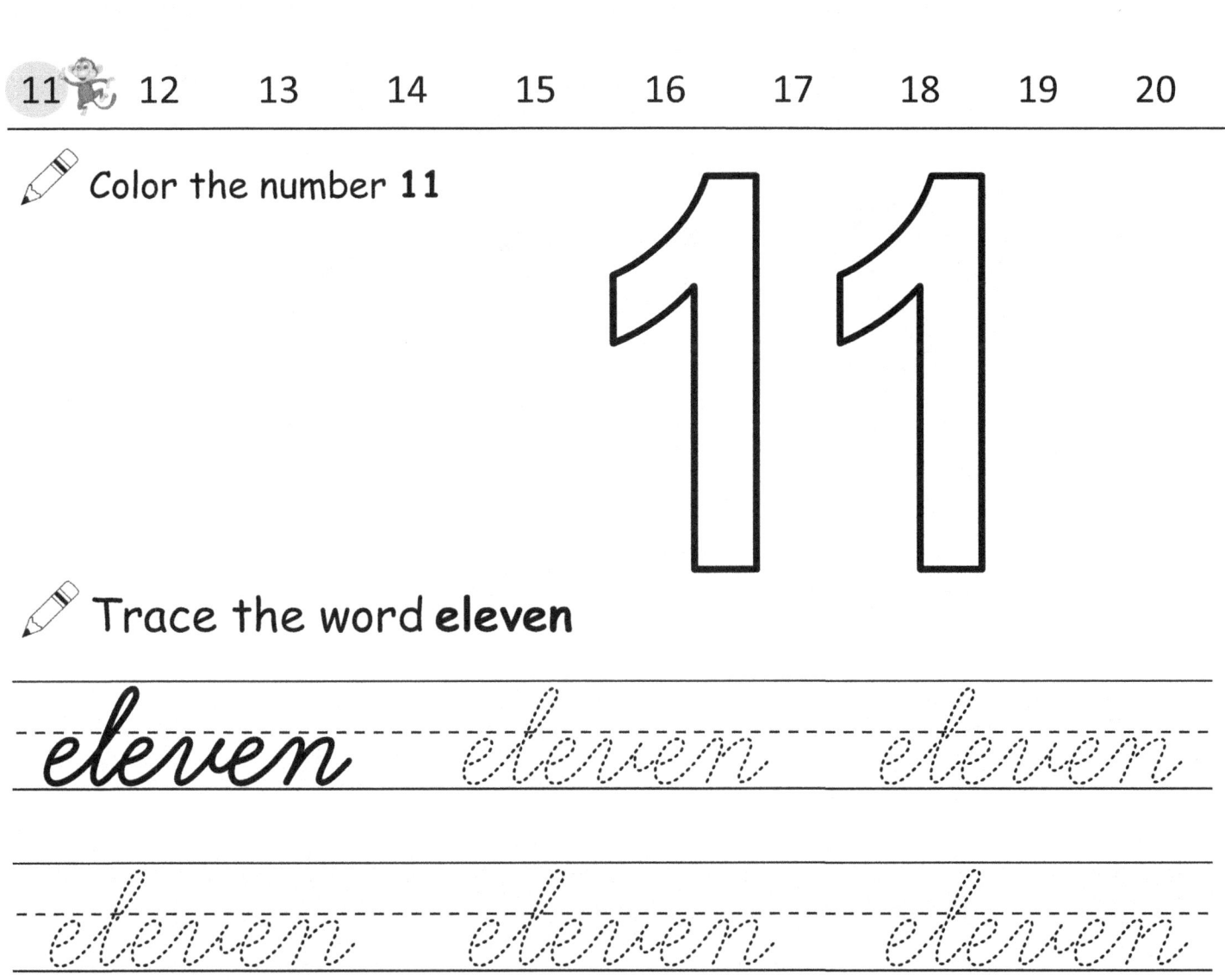

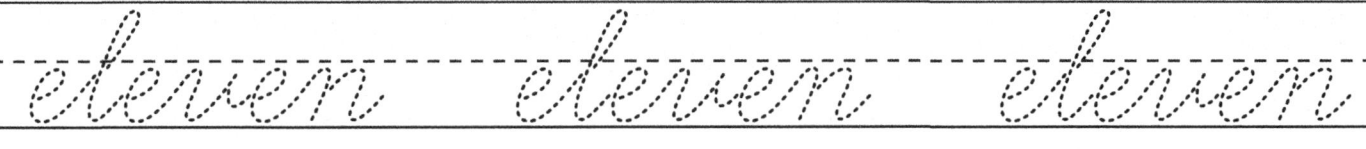

11 12 13 14 15 16 17 18 19 20

✏️ Color the number 11

11

✏️ Trace the word **eleven**

eleven   *eleven*   *eleven*

eleven   *eleven*   *eleven*

11 Trees

11 12 13 14 15 16 17 18 19 20

12 twelve

✏️ **Trace the number 12**

12 12 12 12 12 12 12

12 12 12 12 12 12 12

12 12 12 12 12 12 12

12 12 12 12 12 12 12

12 12 12 12 12 12 12

11   12   13   14   15   16   17   18   19   20

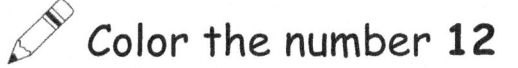

✏️ Color the number **12**

12

✏️ Trace the word **twelve**

twelve   twelve   twelve

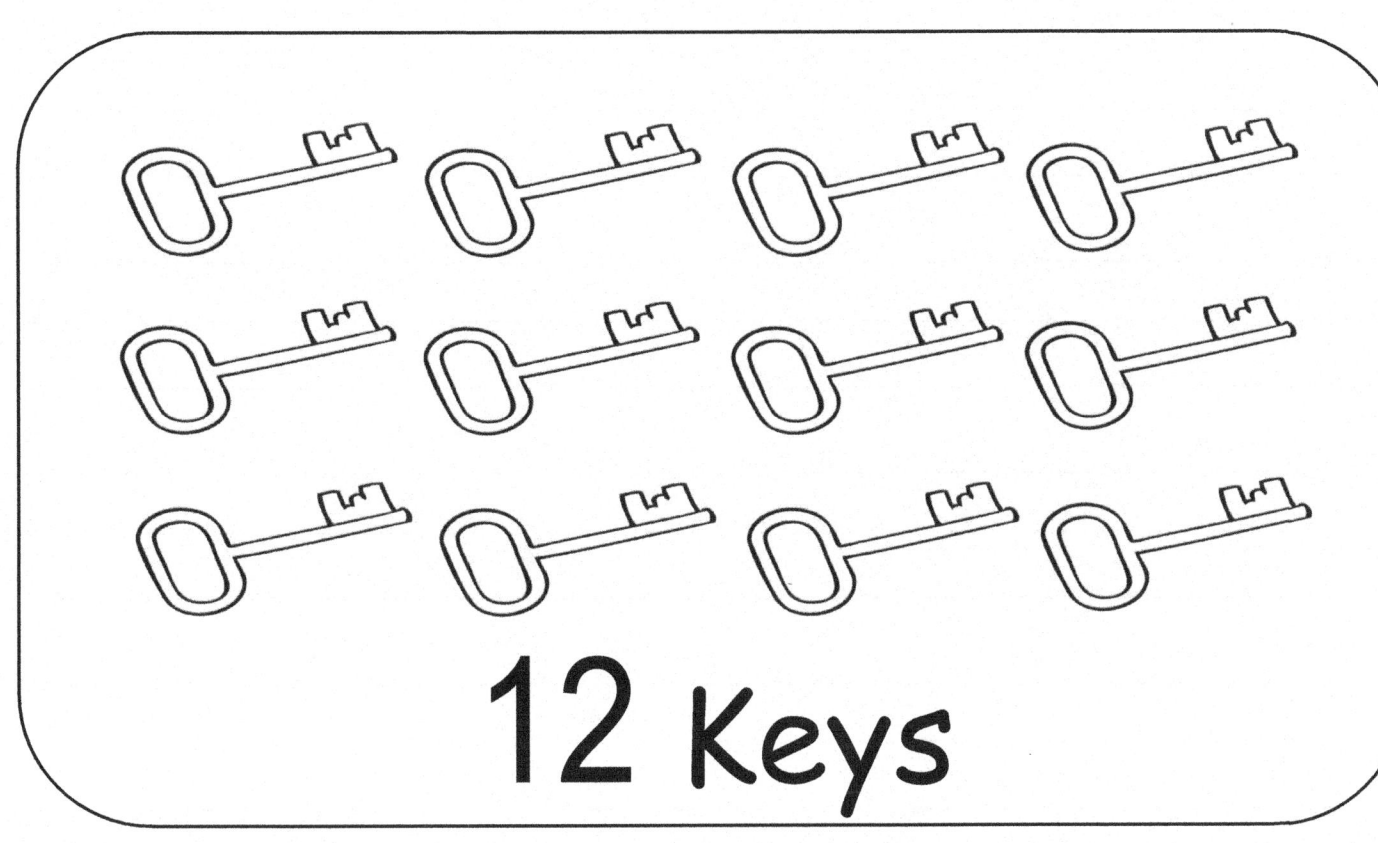

twelve   twelve   twelve

12 Keys

11   12   **13**   14   15   16   17   18   19   20

13   thirteen

✏️ Trace the number **13**

13 13 13 13 13 13

13 13 13 13 13 13

13 13 13 13 13 13

13 13 13 13 13 13

13 13 13 13 13 13

11 12 13 14 15 16 17 18 19 20

✏️ Color the number **13**

13

✏️ Trace the word **thirteen**

thirteen   *thirteen*

thirteen   *thirteen*

13 Strawberries

11　　12　　13　　14　　15　　16　　17　　18　　19　　20

14 fourteen

✏️ Trace the number **14**

14　14　14　14　14　14

14　14　14　14　14　14

14　14　14　14　14　14

14　14　14　14　14　14

14　14　14　14　14　14

11 12 13 **14** 15 16 17 18 19 20

✏️ Color the number **14**

14

✏️ Trace the word **fourteen**

fourteen   *fourteen*

fourteen   *fourteen*

14 Hats

11　　12　　13　　14　　15　　16　　17　　18　　19　　20

15 fifteen

✏️ **Trace the number 15**

15　15　15　15　15　15

15　15　15　15　15　15

15　15　15　15　15　15

15　15　15　15　15　15

15　15　15　15　15　15

11   12   13   14   **15**   16   17   18   19   20

✏️ Color the number **15**

15

✏️ Trace the word **fifteen**

fifteen   *fifteen*   *fifteen*

fifteen   *fifteen*   *fifteen*

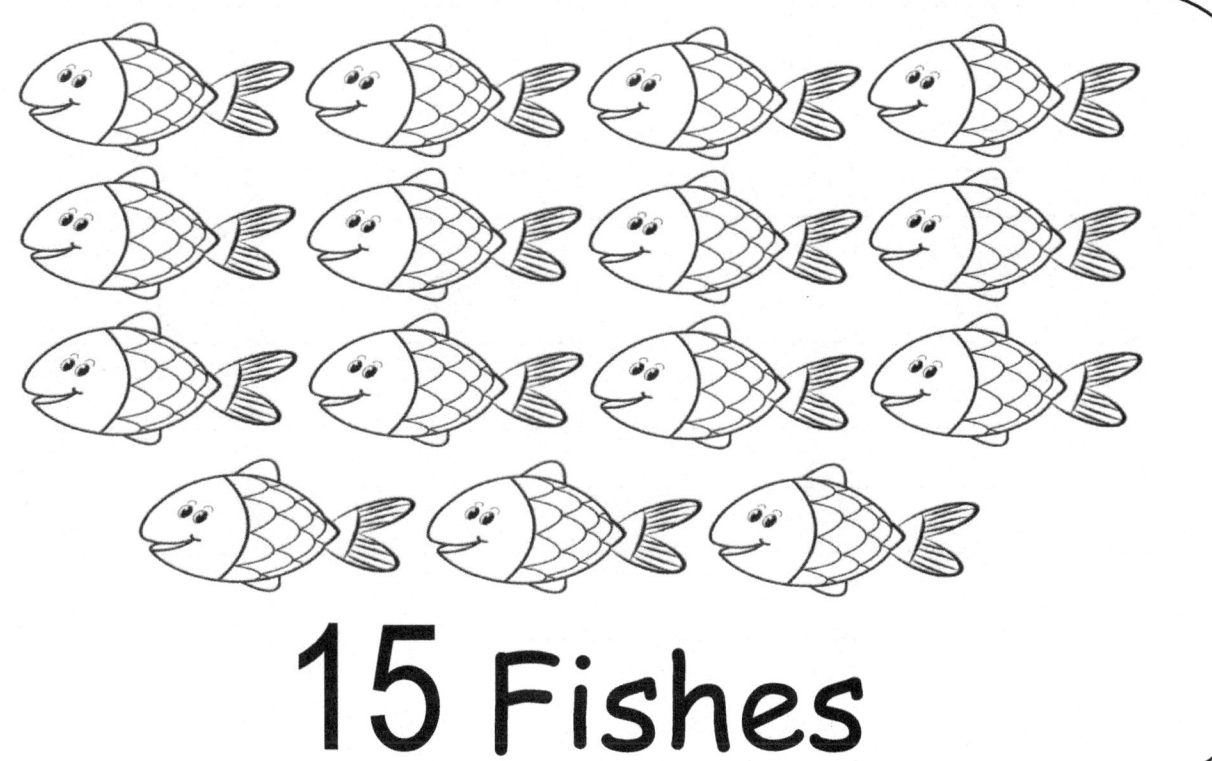

15 Fishes

11   12   13   14   15   16   17   18   19   20

16 sixteen

✏️ Trace the number **16**

16 16 16 16 16 16
16 16 16 16 16 16
16 16 16 16 16 16
16 16 16 16 16 16
16 16 16 16 16 16

11   12   13   14   15   16   17   18   19   20

✏️ Color the number **16**

16

✏️ Trace the word **sixteen**

sixteen   *sixteen*   *sixteen*

sixteen   *sixteen*   *sixteen*

16 Balls

11 12 13 14 15 16 **17** 18 19 20

17 seventeen

✏️ Trace the number **17**

17 17 17 17 17 17

17 17 17 17 17 17

17 17 17 17 17 17

17 17 17 17 17 17

17 17 17 17 17 17

11　　12　　13　　14　　15　　16　　17　　18　　19　　20

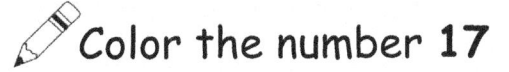

✏️ Color the number **17**

17

✏️ Trace the word **seventeen**

seventeen　　*seventeen*

seventeen　　*seventeen*

17 Bees

11　　12　　13　　14　　15　　16　　17　　**18**　　19　　20

18 eighteen

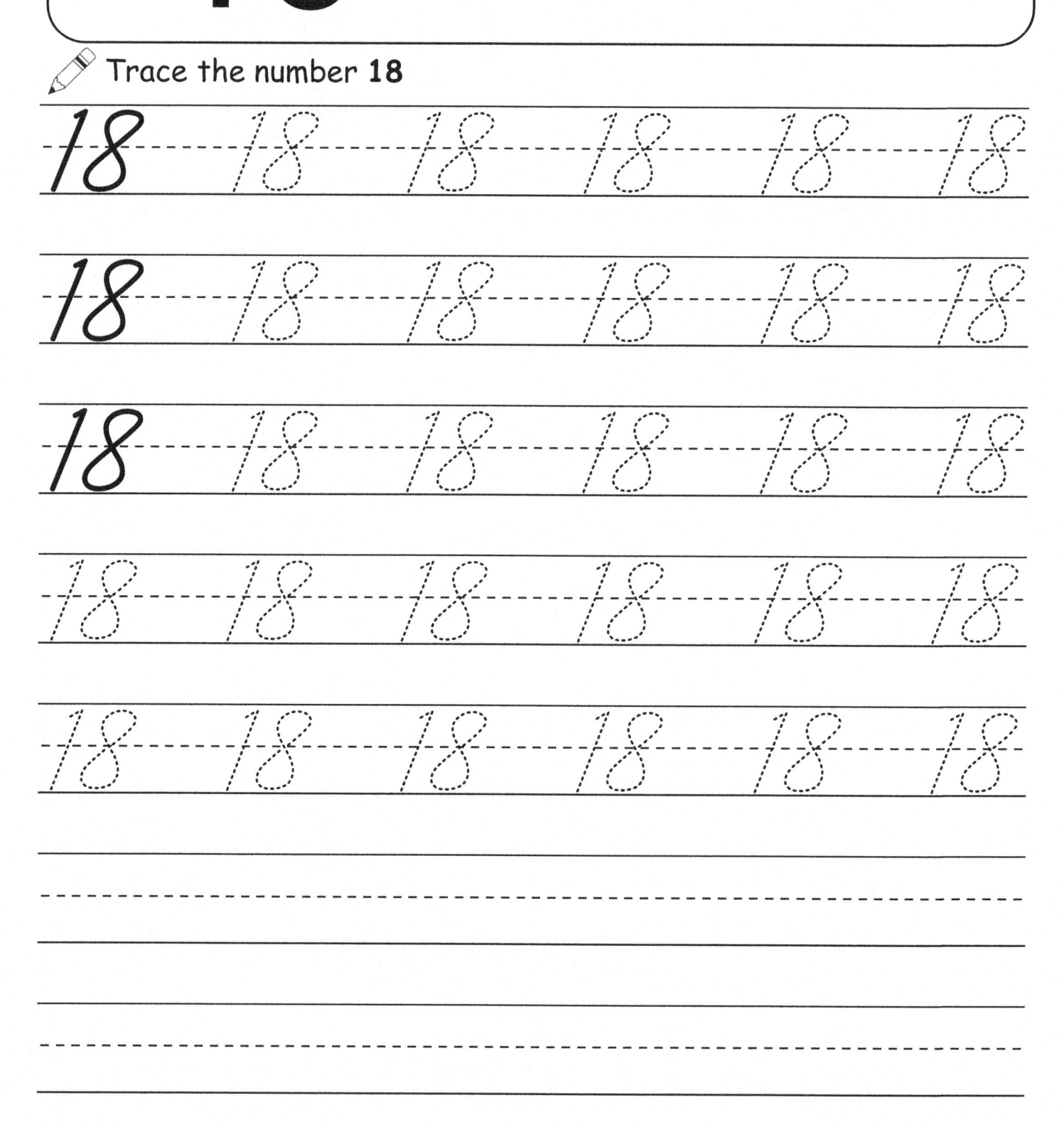

✏️ Trace the number **18**

18　18　18　18　18　18

18　18　18　18　18　18

18　18　18　18　18　18

18　18　18　18　18　18

18　18　18　18　18　18

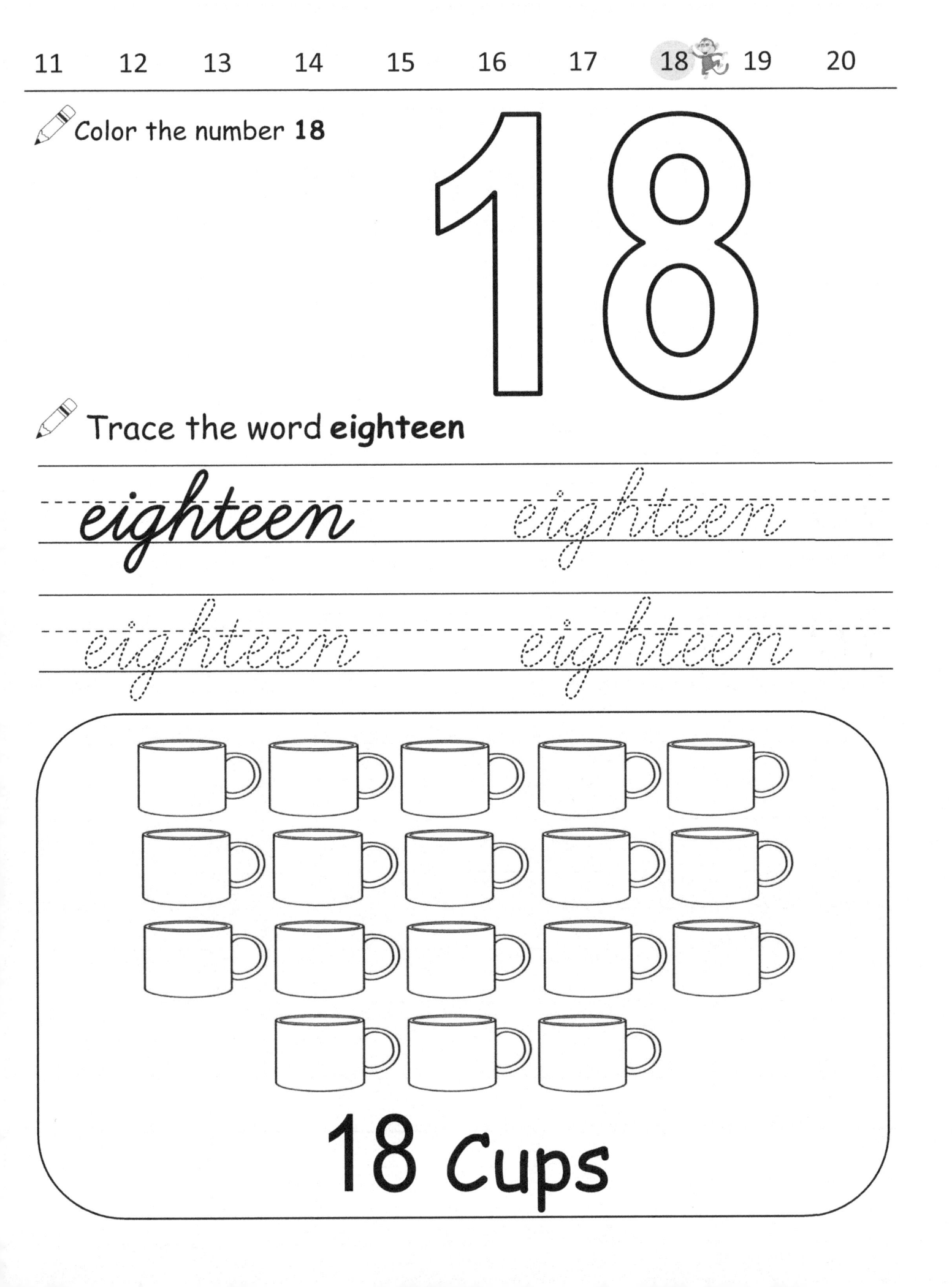

11   12   13   14   15   16   17   18   19   20

✏️ Color the number **18**

18

✏️ Trace the word **eighteen**

eighteen   *eighteen*

eighteen   *eighteen*

18 Cups

11 12 13 14 15 16 17 18 19 20

19 nineteen

✏️ Trace the number **19**

19 19 19 19 19 19
19 19 19 19 19 19
19 19 19 19 19 19
19 19 19 19 19 19
19 19 19 19 19 19

11 12 13 14 15 16 17 18 19 20

✏️ Color the number **19**

19

✏️ Trace the word **nineteen**

nineteen   *nineteen*

nineteen   *nineteen*

19 Stars

11 12 13 14 15 16 17 18 19 20

20 twenty

✏️ Trace the number **20**

20 20 20 20 20 20

20 20 20 20 20 20

20 20 20 20 20 20

20 20 20 20 20 20

20 20 20 20 20 20

11 12 13 14 15 16 17 18 19 20

✏️ Color the number **20**

20

✏️ Trace the word **twenty**

twenty *twenty* *twenty*

twenty *twenty* *twenty*

20 Rabbits

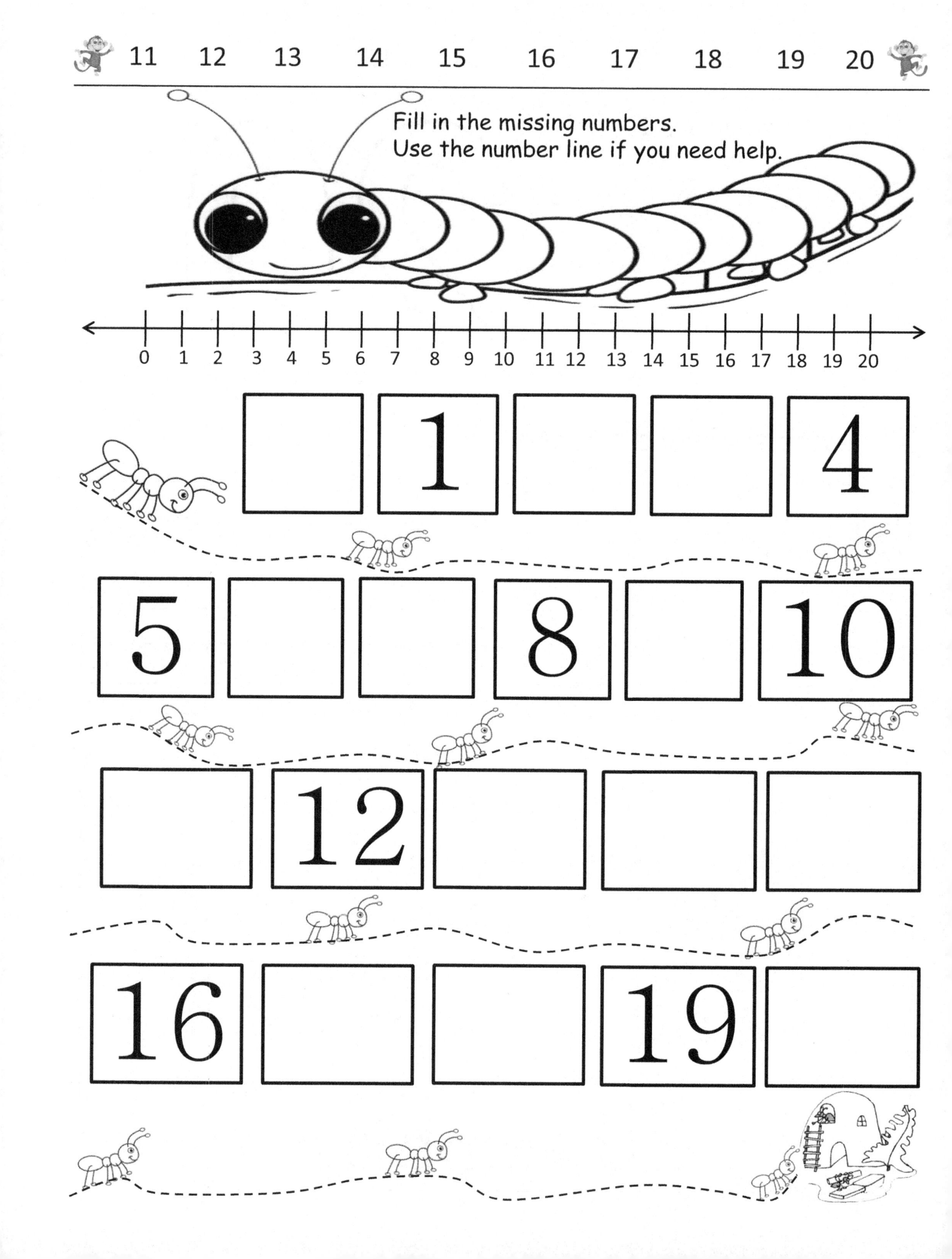

11 12 13 14 15 16 17 18 19 20

Fill in the missing numbers.
Use the number line if you need help.

0 1 2 3 4 5 6 7 8 9 10 11 12 13 14 15 16 17 18 19 20

| | 1 | | | 4 |

| 5 | | | 8 | | 10 |

| | 12 | | | |

| 16 | | | 19 | |

11 12 13 14 15 16 17 18 19 20

✏️ Count the objects and write the number in the box.

1) = ☐

2) = ☐

3) = ☐

4) = ☐

5) = ☐

6) = ☐

7) = ☐

8) = ☐

Number Bonds

💡 Discover

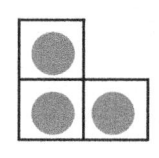

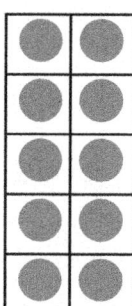

3 + ? = 10

Part   Part   Whole

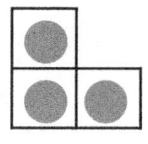

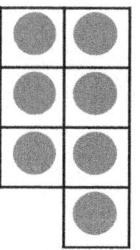

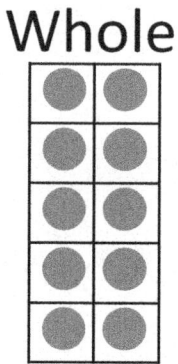

3 + 7 = 10

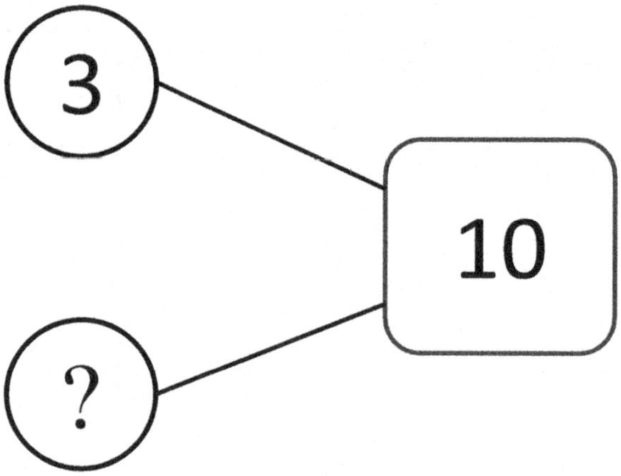

3 — 10
?

3 + ? = 10

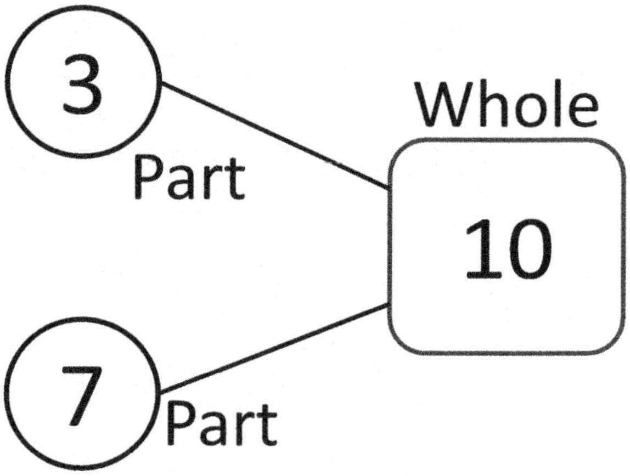

3 Part — Whole 10
7 Part

3 + 7 = 10

Number Bonds

✏️ Draw the missing objects to make a total of 10.

6 + 4 = 10

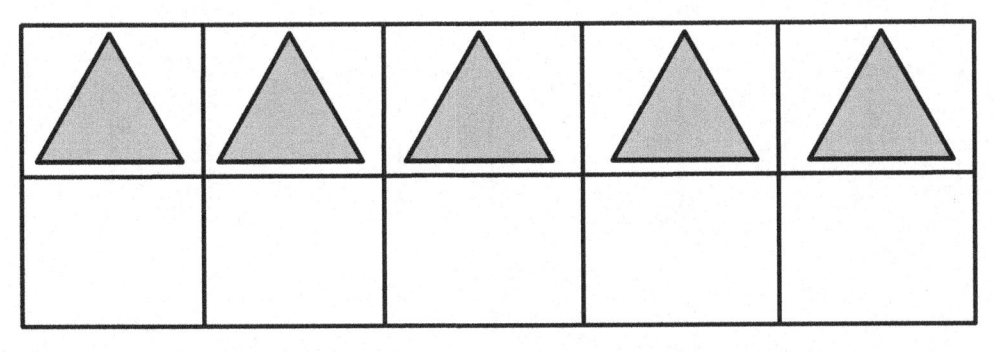

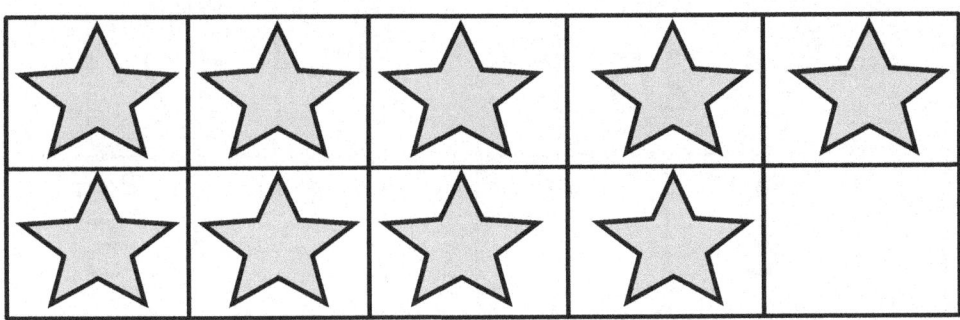

8 + ___ = 10

3 + ___ = 10

5 + ___ = 10

9 + ___ = 10

Number Bonds

✏️ What's missing number here?

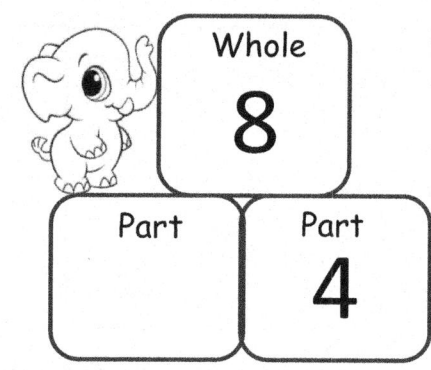

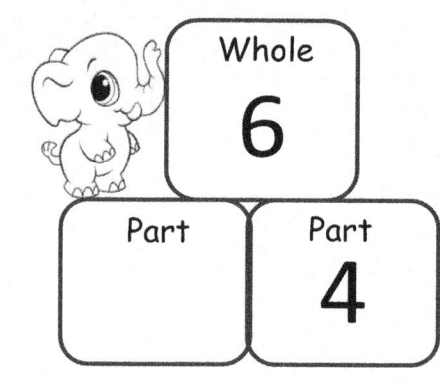

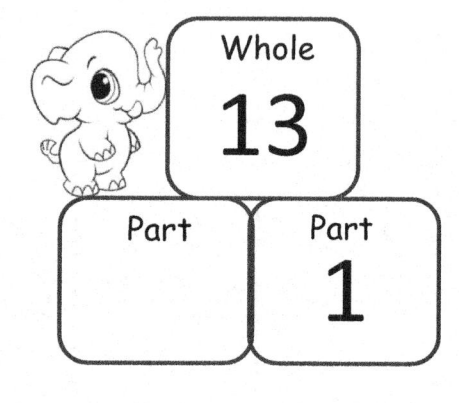

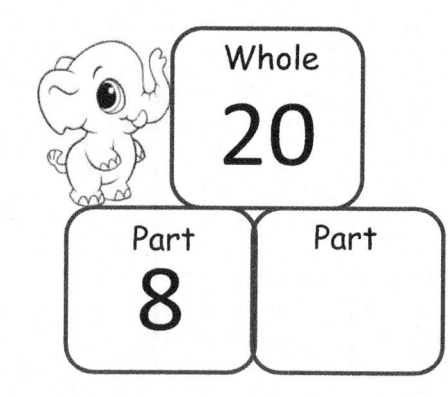

Whole	Whole	Whole
3	5	8
Part 2 / Part 1	Part 2 / Part	Part / Part 4

Whole	Whole	Whole
4	6	9
Part 1 / Part	Part / Part 4	Part 7 / Part

Whole	Whole	Whole
11	16	13
Part 5 / Part	Part 3 / Part	Part / Part 1

Whole	Whole	Whole
15	20	17
Part 2 / Part	Part 8 / Part	Part / Part 5

Number Bonds

✏️ Complete the number bond. Finish the number sentence to match.

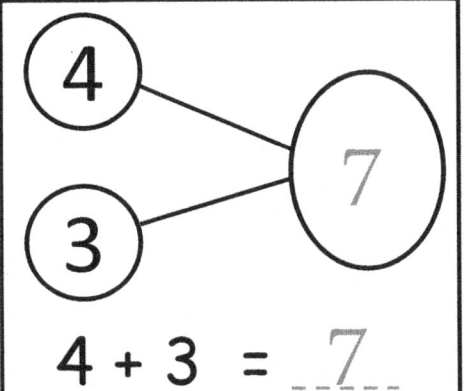

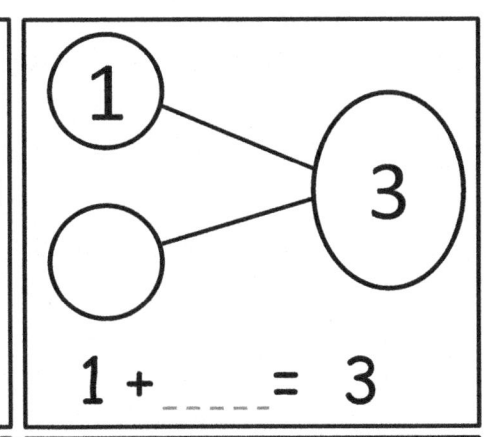

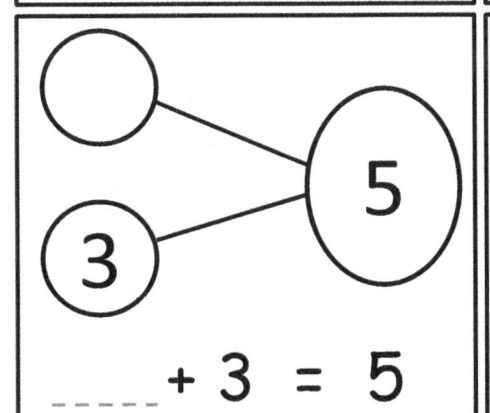

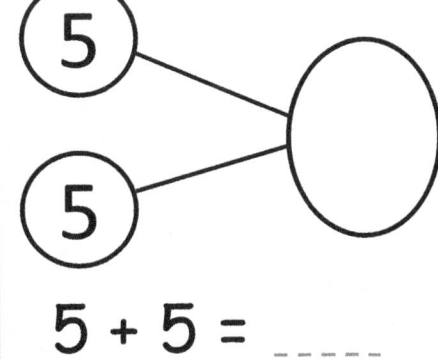

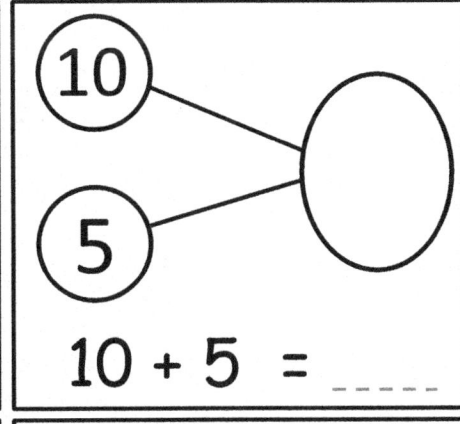

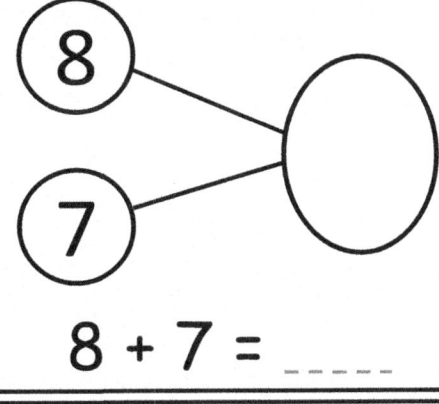

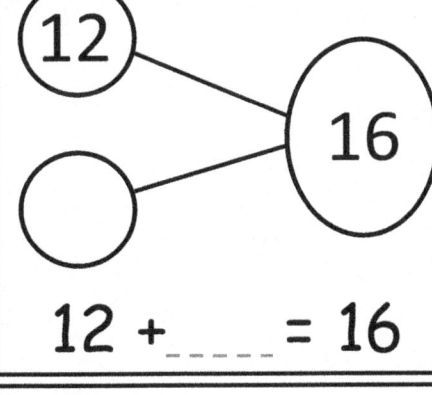

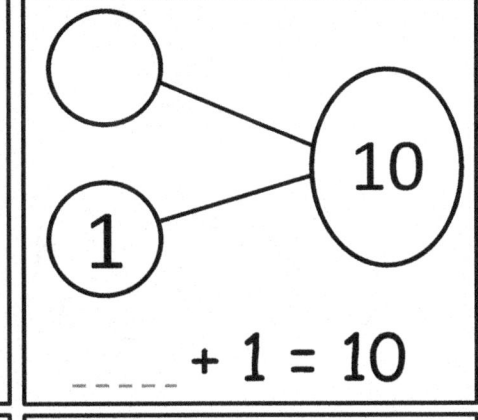

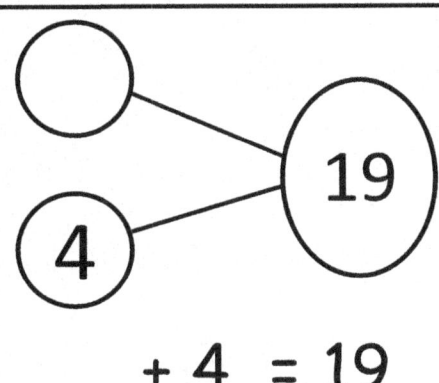

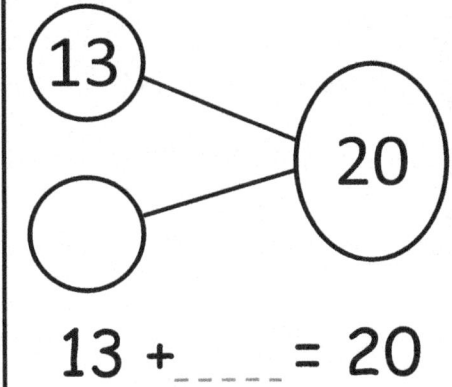

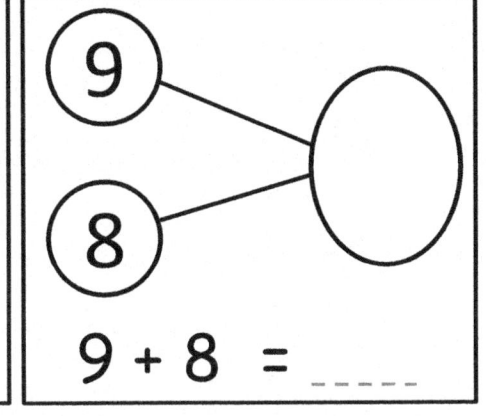

4, 3, 7 4 + 3 = _7_	4, 5, ◯ 4 + 5 = ____	1, ◯, 3 1 + ____ = 3
◯, 3, 5 ____ + 3 = 5	5, 5, ◯ 5 + 5 = ____	10, 5, ◯ 10 + 5 = ____
8, 7, ◯ 8 + 7 = ____	12, ◯, 16 12 + ____ = 16	◯, 1, 10 ____ + 1 = 10
◯, 4, 19 ____ + 4 = 19	13, ◯, 20 13 + ____ = 20	9, 8, ◯ 9 + 8 = ____

Addition

💡 Discover

[2] + [1] = [3]

[2] + [2] = [4]

[2] + [3] = [5]

Addition

✏️ Count and write the correct number in the box.

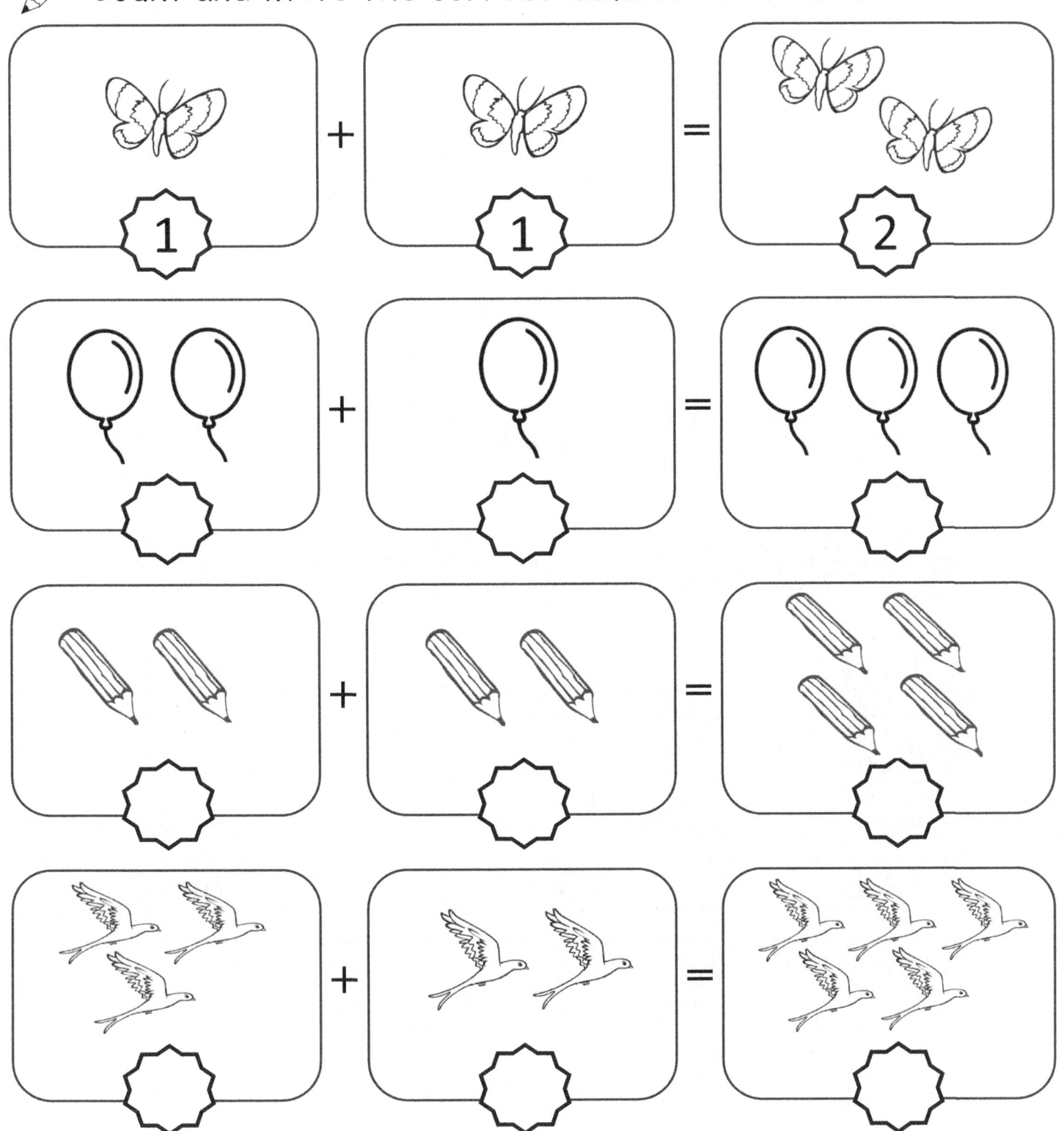

1 + 1 = 2

Addition

Circle the correct sum that matches the picture.

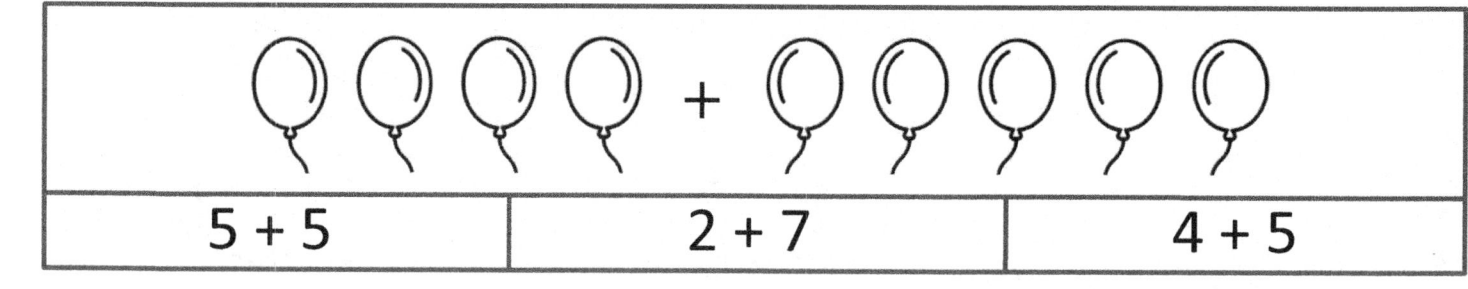

| 3 + 5 | 2 + 4 | (2 + 3) |

| 5 + 1 | 4 + 3 | 1 + 6 |

| 4 + 4 | 3 + 7 | 6 + 5 |

| 2 + 8 | 3 + 4 | 6 + 4 |

| 5 + 5 | 2 + 7 | 4 + 5 |

Addition

✏️ Count, sum and circle the correct number.

☆☆☆☆ + ☆☆ = 4 5 6

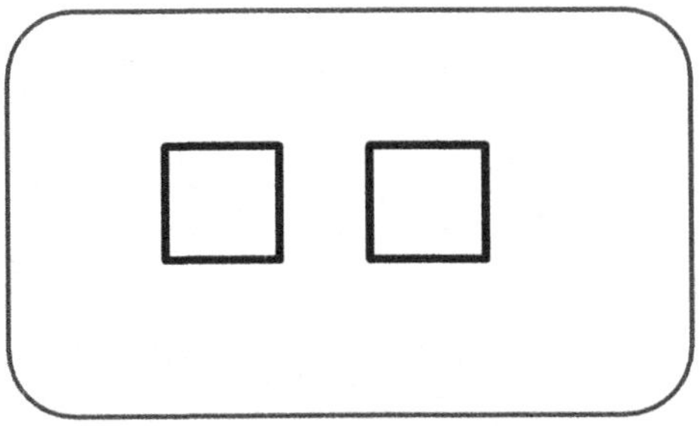

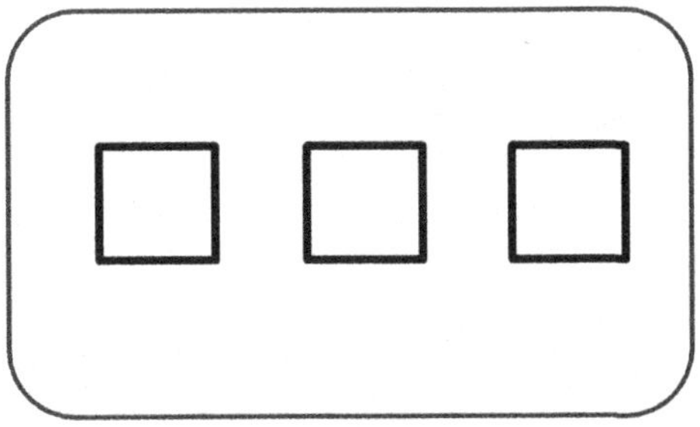

□□ + □□□ = 5 6 7

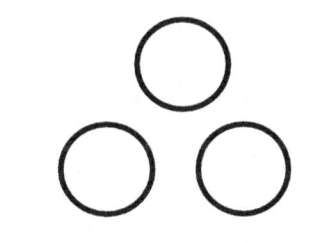

○○○○○ + ○○○ = 6 7 8

♡ + ♡♡♡♡♡♡ = 5 6 7

△△△ + △△△ = 4 5 6

Addition

✏️ Count and add

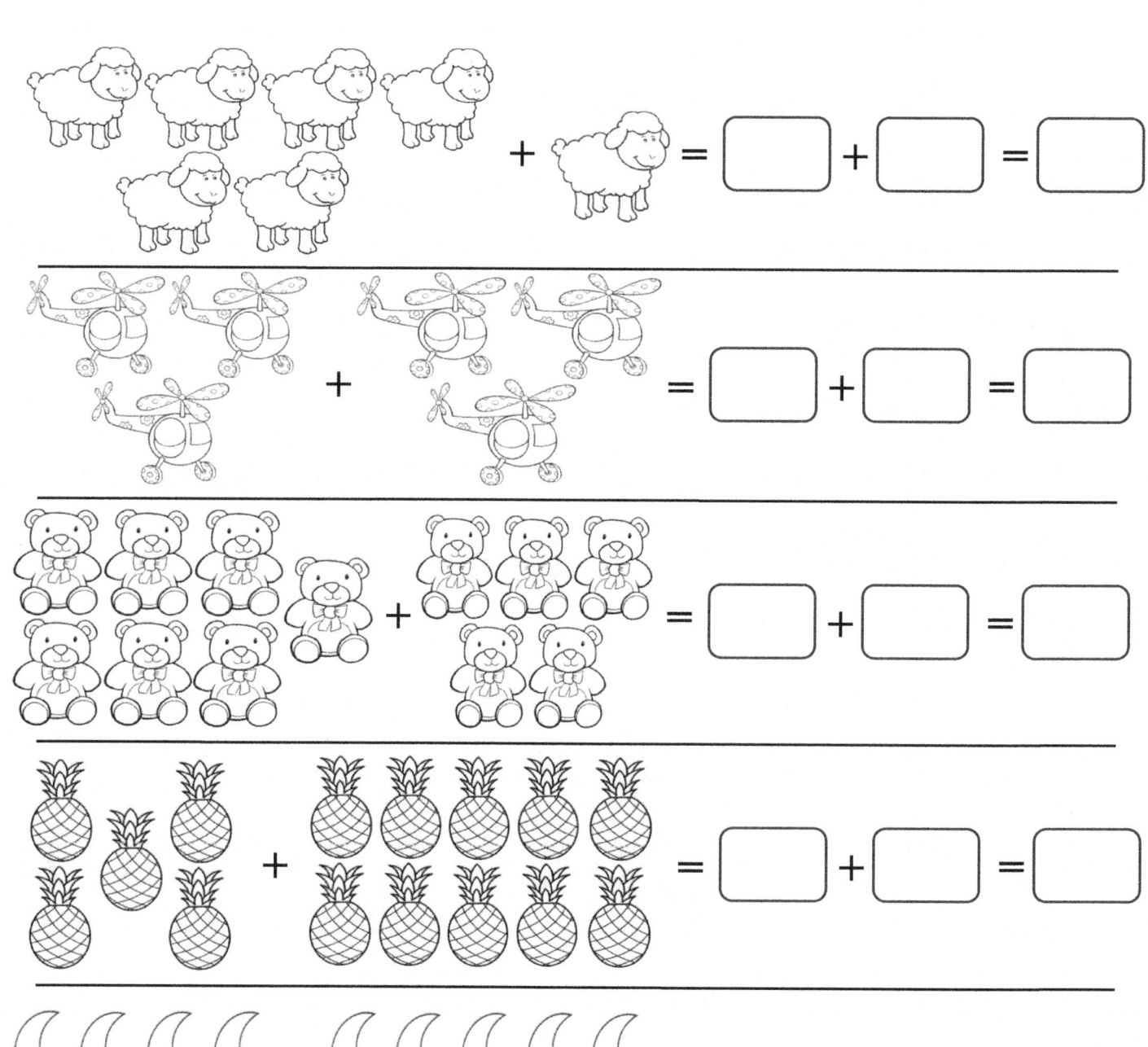

6 + 1 = ☐ + ☐ = ☐

4 + 3 = ☐ + ☐ = ☐

7 + 5 = ☐ + ☐ = ☐

6 + 9 = ☐ + ☐ = ☐

8 + 8 = ☐ + ☐ = ☐

Addition

✏️ Count and write the correct number in the box.

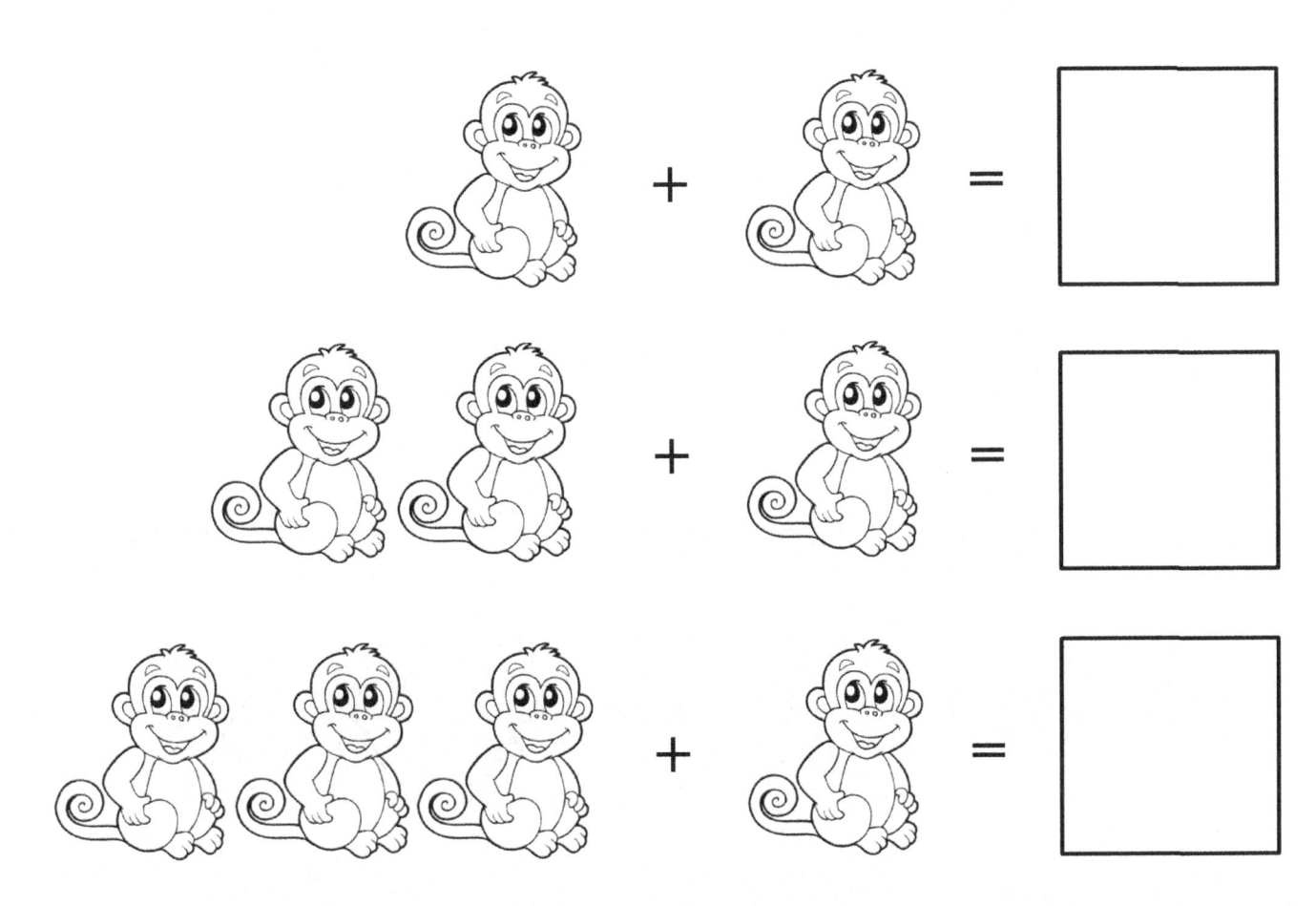

🐵 + 🐵 = ☐

🐵🐵 + 🐵 = ☐

🐵🐵🐵 + 🐵 = ☐

- -

1 + 1 = ☐   3 + 1 = ☐

2 + 1 = ☐   4 + 1 = ☐

Subtraction

💡 Discover

$4 - 1 = 3$

$7 - 3 = 4$

$8 - 2 = 6$

Subtraction

✏️ Count the pictures and subtract.

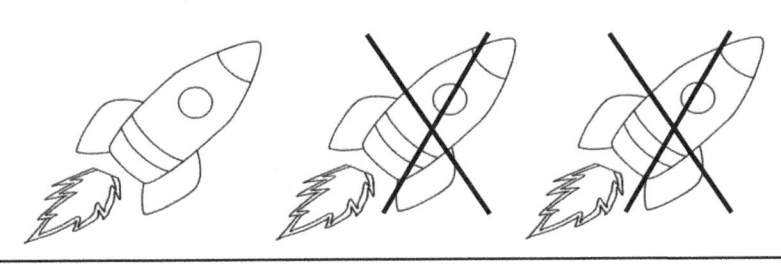

3 − 2 = 1

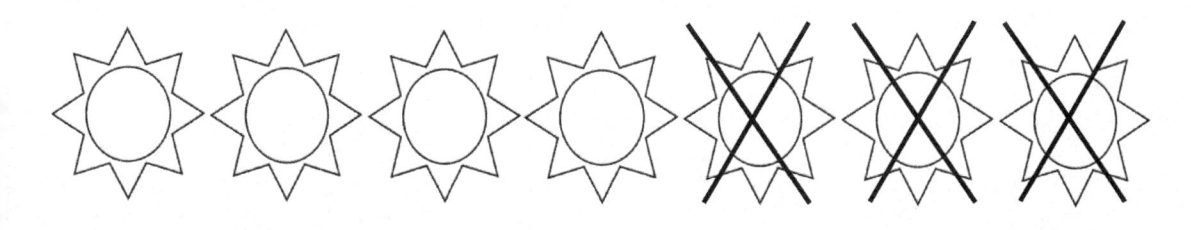

6 − 4 = ____

4 − 1 = ____

7 − 3 = ____

5 − 2 = ____

8 − 5 = ____

Subtraction

✏️ Write the answer under the line. Use the shapes for help.

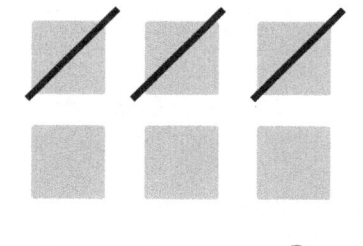

6 − 3 = 3

5 − 2 = ___

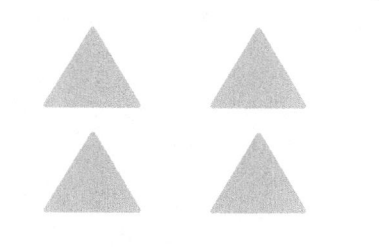

4 − 1 = ___

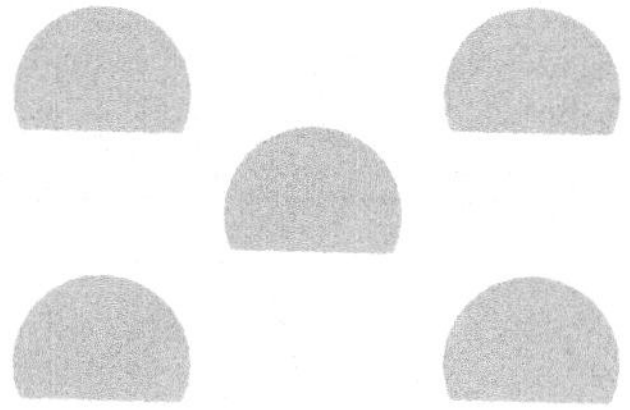

5 − 4 = ___

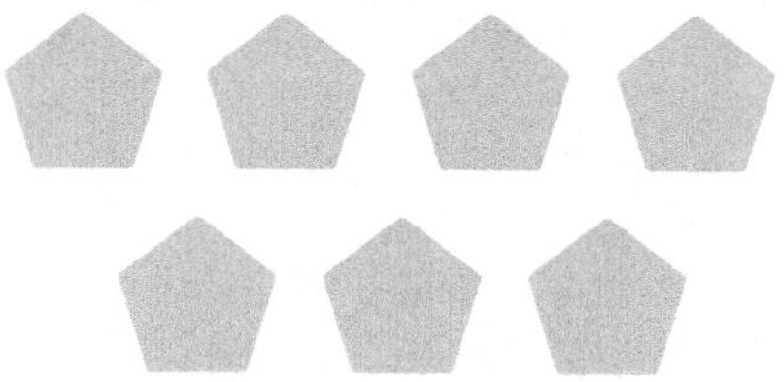

7 − 5 = ___

6 − 2 = ___

8 − 5 = ___

6 − 3 = ___

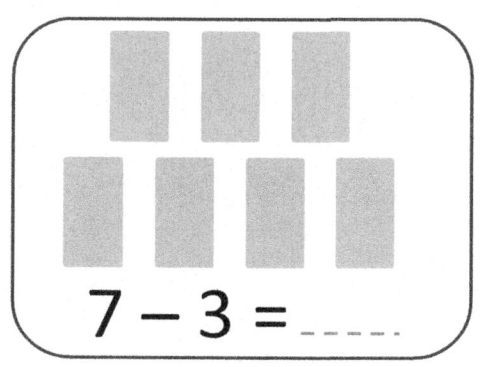

7 − 3 = ___

9 − 4 = ___

7 − 6 = ___

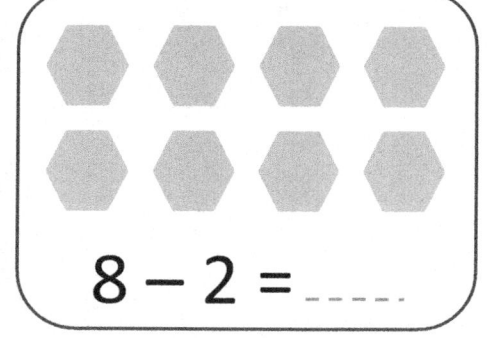

8 − 2 = ___

Subtraction

Subtract to find each difference.

5 − 2 = ☐   4 − 3 = ☐

8 − 4 = ☐   7 − 5 = ☐

6 − 1 = ☐   8 − 2 = ☐

7 − 3 = ☐   5 − 4 = ☐

9 − 3 = ☐   9 − 7 = ☐

Subtraction

✏️ Count, subtract and write the result under the line.

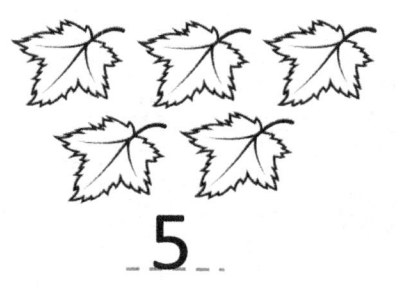

5 − 3 = 2

___ ___ ___

___ − ___ = ___

___ − ___ = ___

___ − ___ = ___

___ − ___ = ___

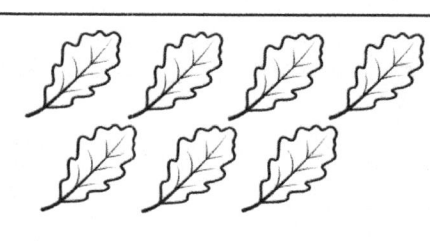

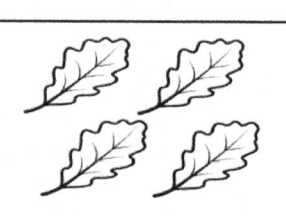

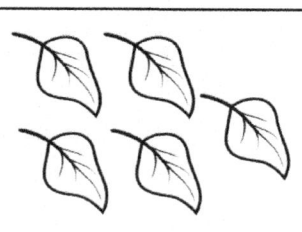

Subtraction

✏️ Count, subtract and circle the correct number.

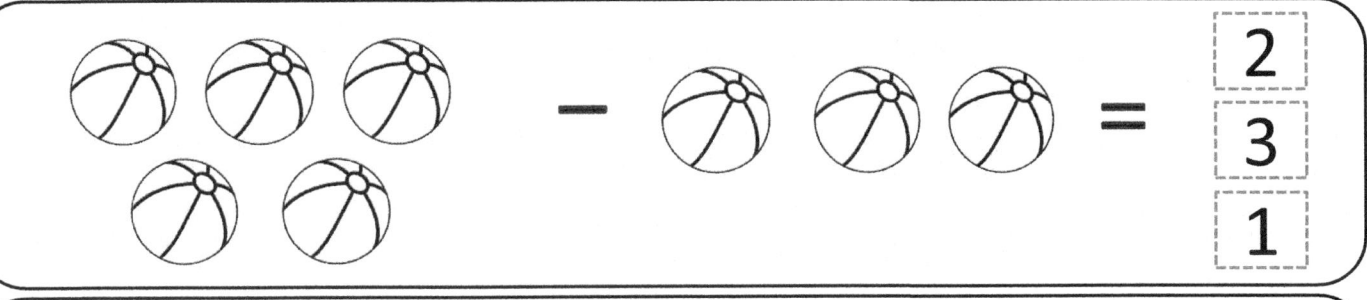

5 - 3 = | 2 / 3 / 1

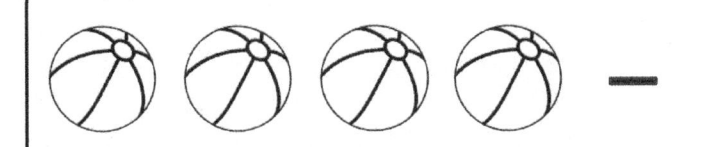

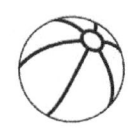

4 - 1 = | 4 / 2 / 3

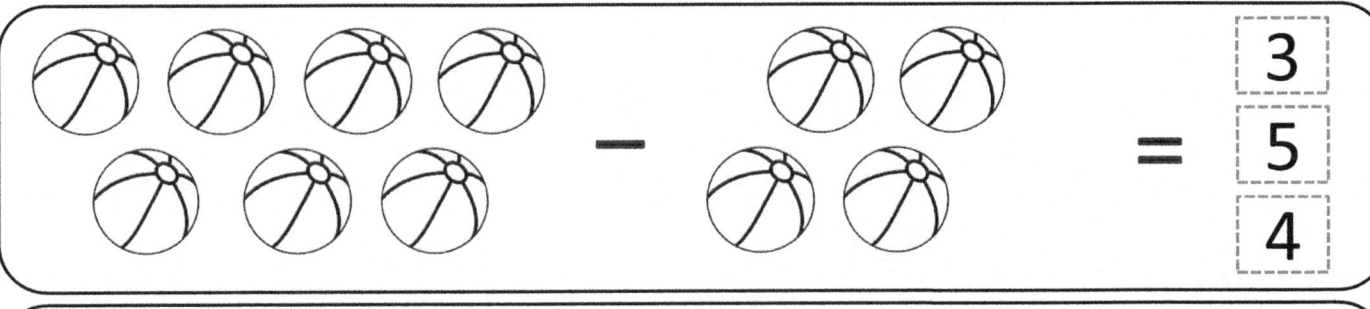

7 - 3 = | 3 / 5 / 4

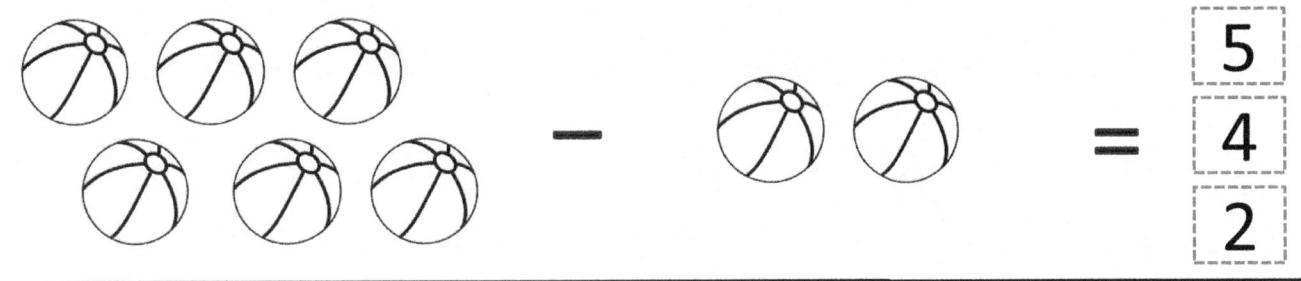

6 - 2 = | 5 / 4 / 2

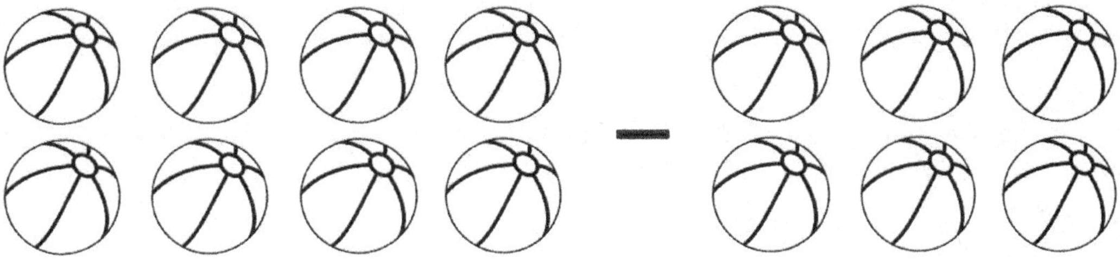

8 - 3 = | 6 / 4 / 2

Subtraction

✏️ Count and subtract

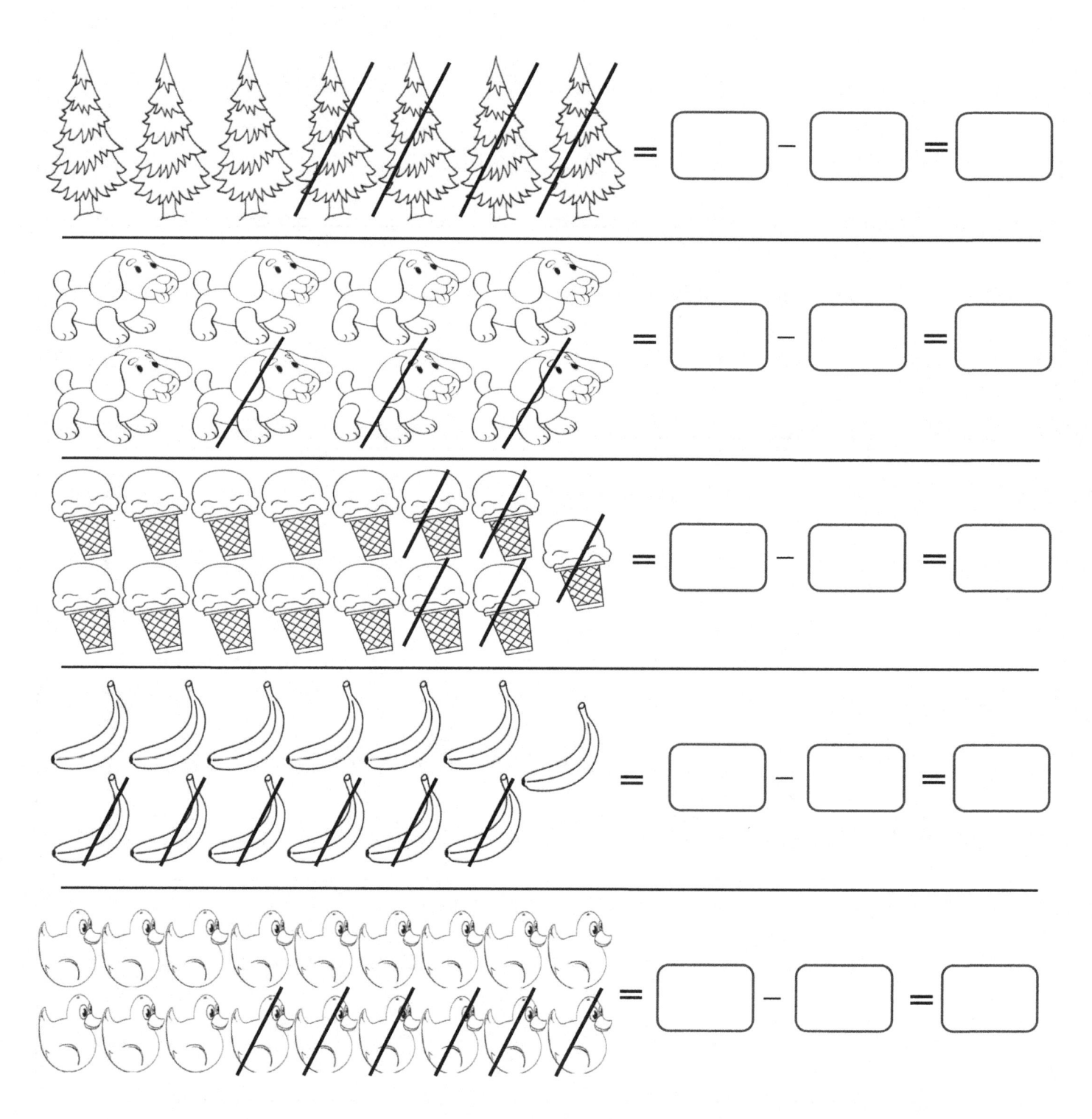

= ☐ - ☐ = ☐

= ☐ - ☐ = ☐

= ☐ - ☐ = ☐

= ☐ - ☐ = ☐

= ☐ - ☐ = ☐

Subtraction

✏️ Count and write the correct number in the box.

4 - 1 = ☐

5 - 1 = ☐

7 - 1 = ☐

- -

4 - 1 = ☐   6 - 1 = ☐

5 - 1 = ☐   7 - 1 = ☐

Addition and Subtraction

I can add...

3 + 4 = -----
1 + 6 = -----
5 + 3 = -----
4 + 4 = -----
9 + 5 = -----
2 + 9 = -----
7 + 3 = -----
8 + 4 = -----
6 + 11 = -----
10 + 2 = -----
13 + 1 = -----
1 + 19 = -----
9 + 6 = -----

I can subtract...

6 - 1 = -----
5 - 4 = -----
3 - 3 = -----
7 - 4 = -----
4 - 2 = -----
8 - 3 = -----
6 - 4 = -----
9 - 9 = -----
10 - 5 = -----
14 - 2 = -----
17 - 10 = -----
15 - 11 = -----
20 - 14 = -----

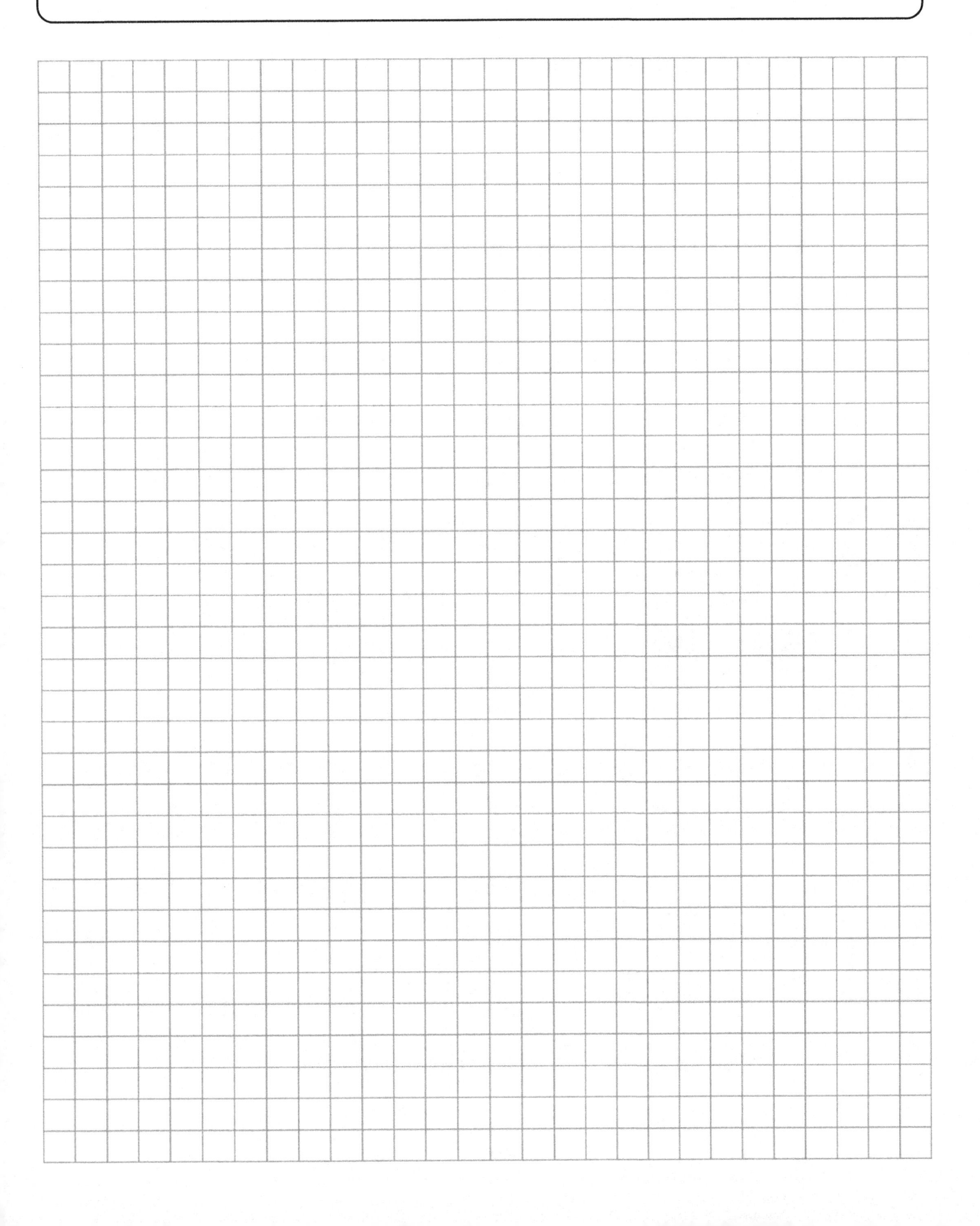

Addition and Subtraction

Addition and Subtraction

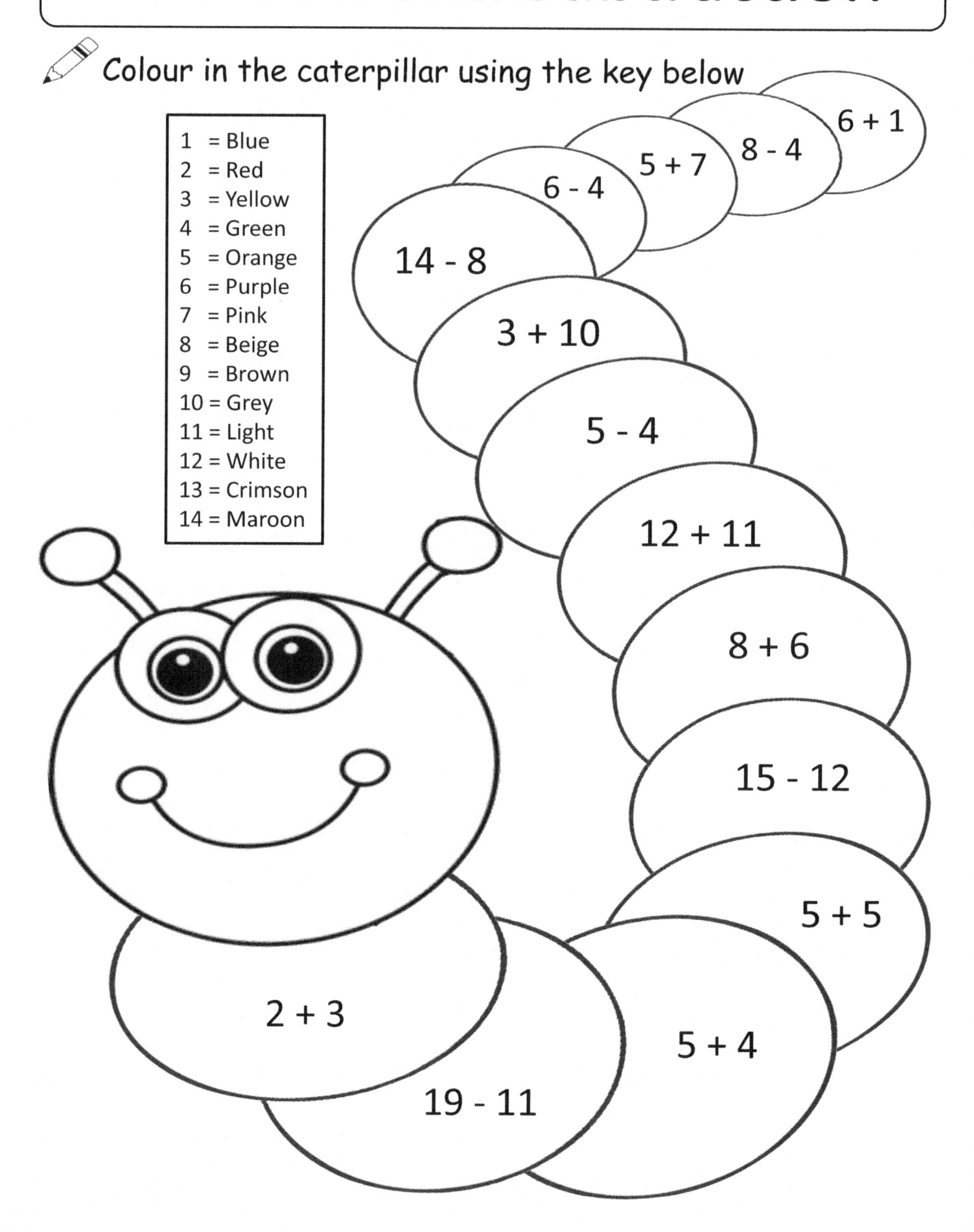

✏️ Colour in the caterpillar using the key below

1 = Blue
2 = Red
3 = Yellow
4 = Green
5 = Orange
6 = Purple
7 = Pink
8 = Beige
9 = Brown
10 = Grey
11 = Light
12 = White
13 = Crimson
14 = Maroon

6 + 1
8 - 4
5 + 7
6 - 4
14 - 8
3 + 10
5 - 4
12 + 11
8 + 6
15 - 12
5 + 5
5 + 4
19 - 11
2 + 3

Numbers to 100

✏️ Trace the numbers to 100

1 - 2 - 3 - 4 - 5 - 6 - 7 - 8 - 9 - 10

11 - 12 - 13 - 14 - 15 - 16 - 17 - 18 - 19 - 20

21 - 22 - 23 - 24 - 25 - 26 - 27 - 28 - 29 - 30

31 - 32 - 33 - 34 - 35 - 36 - 37 - 38 - 39 - 40

41 - 42 - 43 - 44 - 45 - 46 - 47 - 48 - 49 - 50

51 - 52 - 53 - 54 - 55 - 56 - 57 - 58 - 59 - 60

61 - 62 - 63 - 64 - 65 - 66 - 67 - 68 - 69 - 70

71 - 72 - 73 - 74 - 75 - 76 - 77 - 78 - 79 - 80

81 - 82 - 83 - 84 - 85 - 86 - 87 - 88 - 89 - 90

91 - 92 - 93 - 94 - 95 - 96 - 97 - 98 - 99 - 100

Numbers to 100

Trace and read the numbers

10 ten 10 ten 10 ten

20 twenty 20 twenty 20 twenty

30 thirty 30 thirty 30 thirty

40 forty 40 forty 40 forty

50 fifty 50 fifty 50 fifty

60 sixty 60 sixty 60 sixty

70 seventy 70 seventy 70 seventy

80 eighty 80 eighty 80 eighty

90 ninety 90 ninety 90 ninety

100 one hundred 100 one hundred

Numbers to 100

✏️ Write the missing numbers

1	2	3		5	6	7	8		10
11		13	14	15	16	17	18	19	20
21	22	23	24	25			28	29	30
31	32		34	35	36	37	38	39	40
41	42	43	44	45	46		48	49	
51	52	53	54		56	57	58	59	60
	62	63	64	65	66	67	68	69	70
71	72	73		75	76	77		79	80
81	82	83	84	85		87	88	89	90
91	92		94	95	96	97	98		100

Measuring
Size : Big and Small

💡 Discover

Small

Big

Big

Small

Small

Big

Measuring

Size : Small / Big

✏️ Circle the small picture.

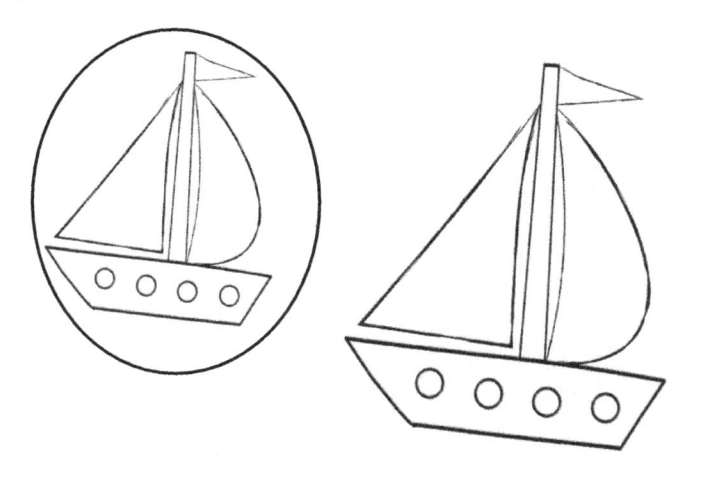

Measuring
Size : Small / Big

✏️ Colour the biggest picture in each row.

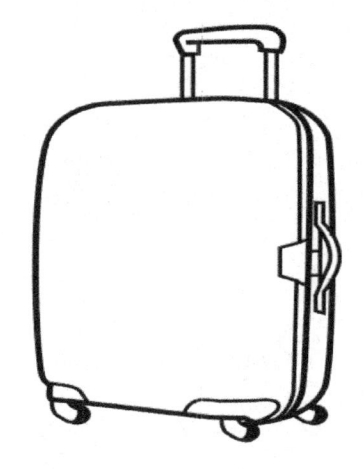

Measuring
Size : Small / Big

✏️ Match the bigger shape on the left to the same smaller shape on the right.

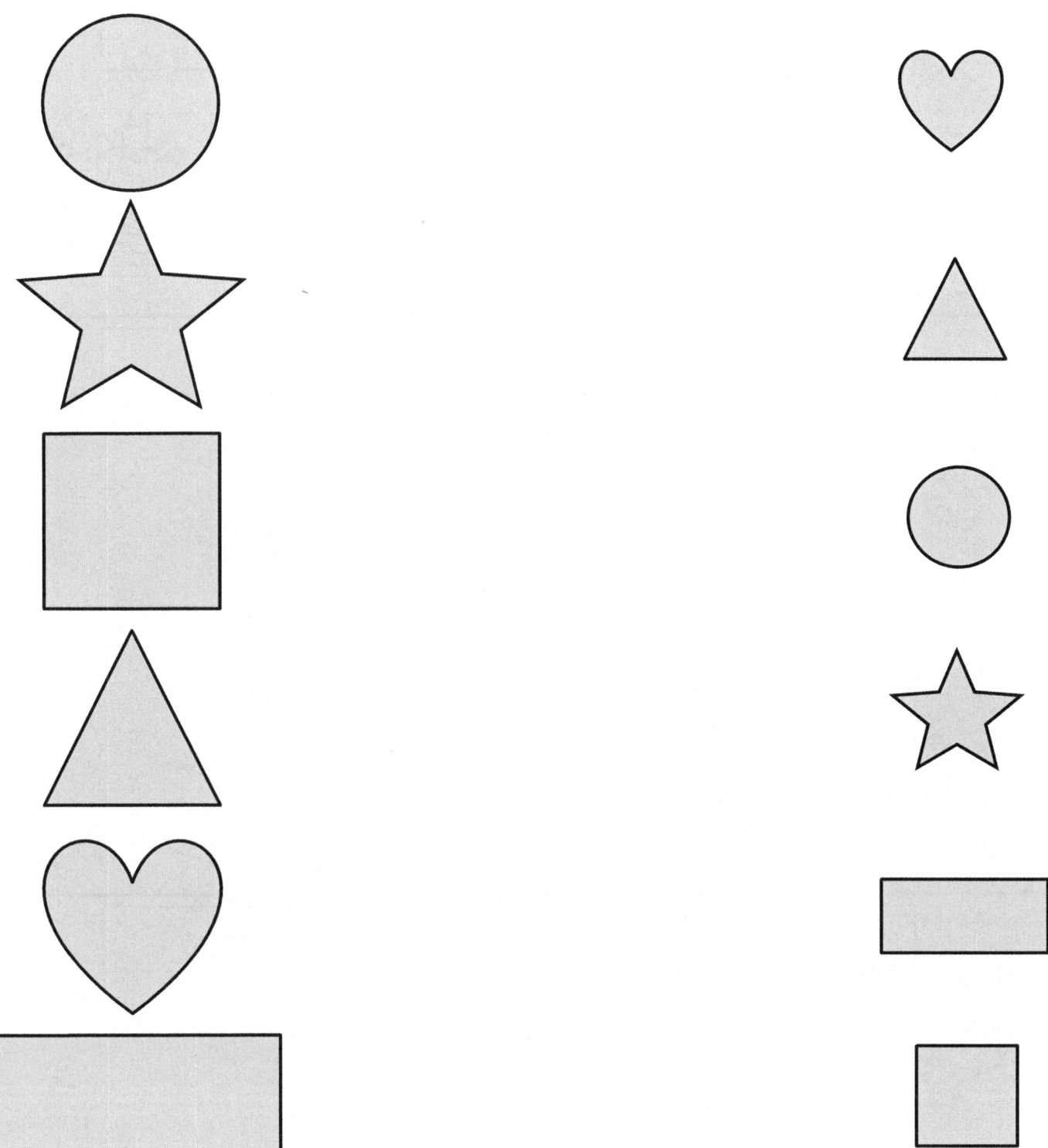

Measuring
Size : Small / Big

Big or Small? Circle the correct picture.

Small	Big
Big	Small
Big	Small
Small	Big

Measuring
Length : Short and Tall

💡 Discover

Short

Tall

Tall

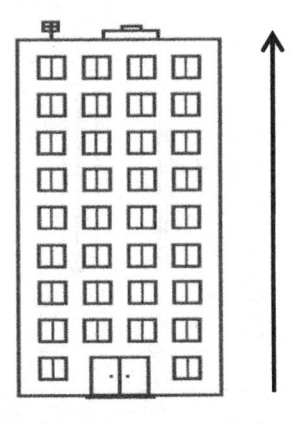

Short

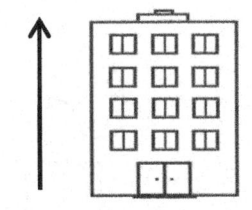

Short

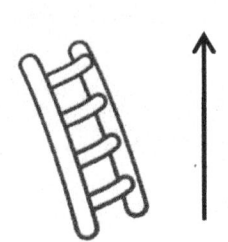

Tall

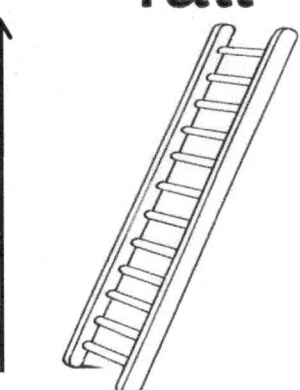

Measuring

Length : Short and Tall

✏️ Which one is taller?

Measuring

Length : Short and Tall

✏️ Colour the shorter picture in each row.

Measuring
Length : Short and Tall

Look at the shapes in each of the six sections.
Color the tall shape red and the short shape yellow.

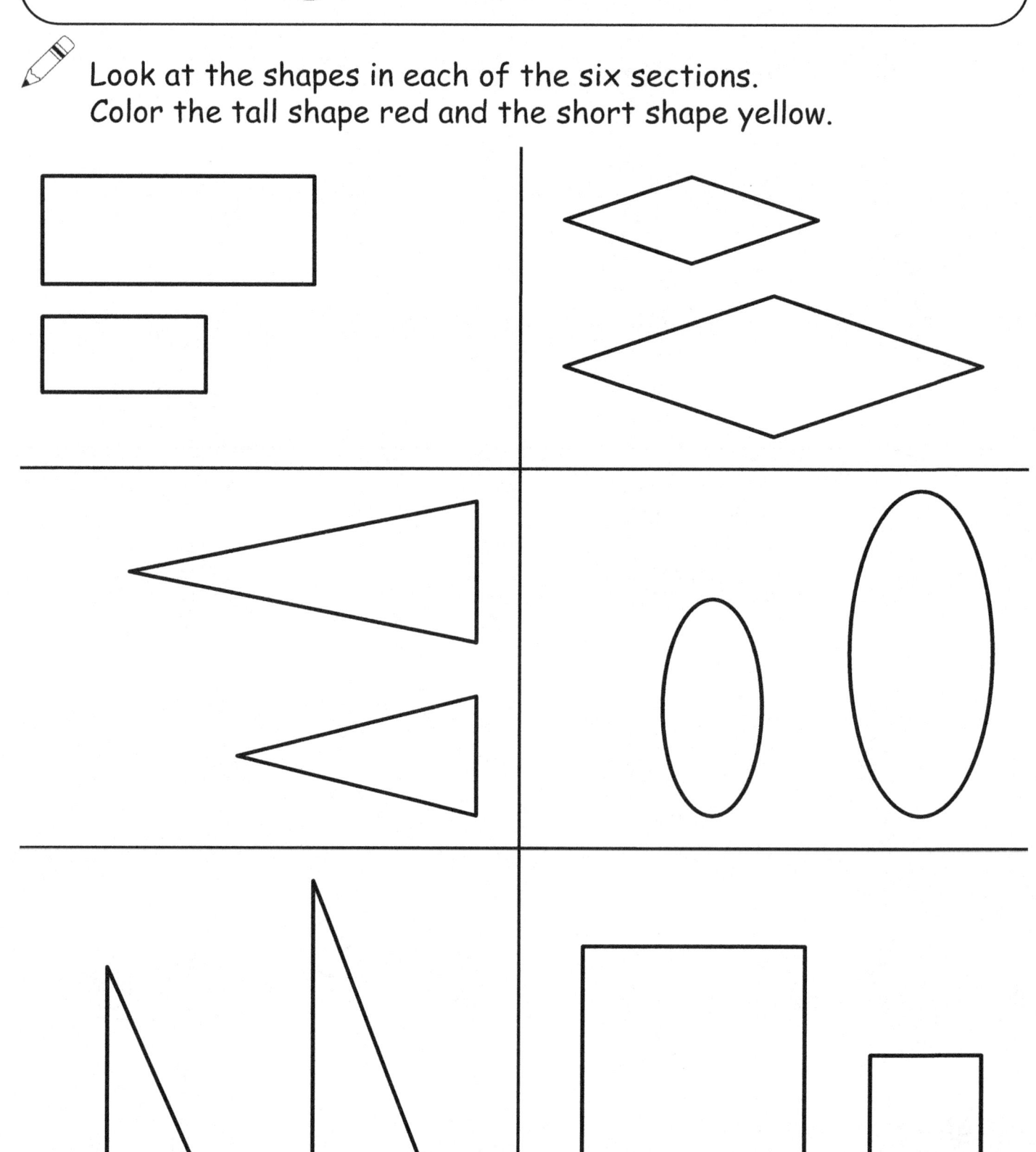

Measuring
Length : Short and Tall

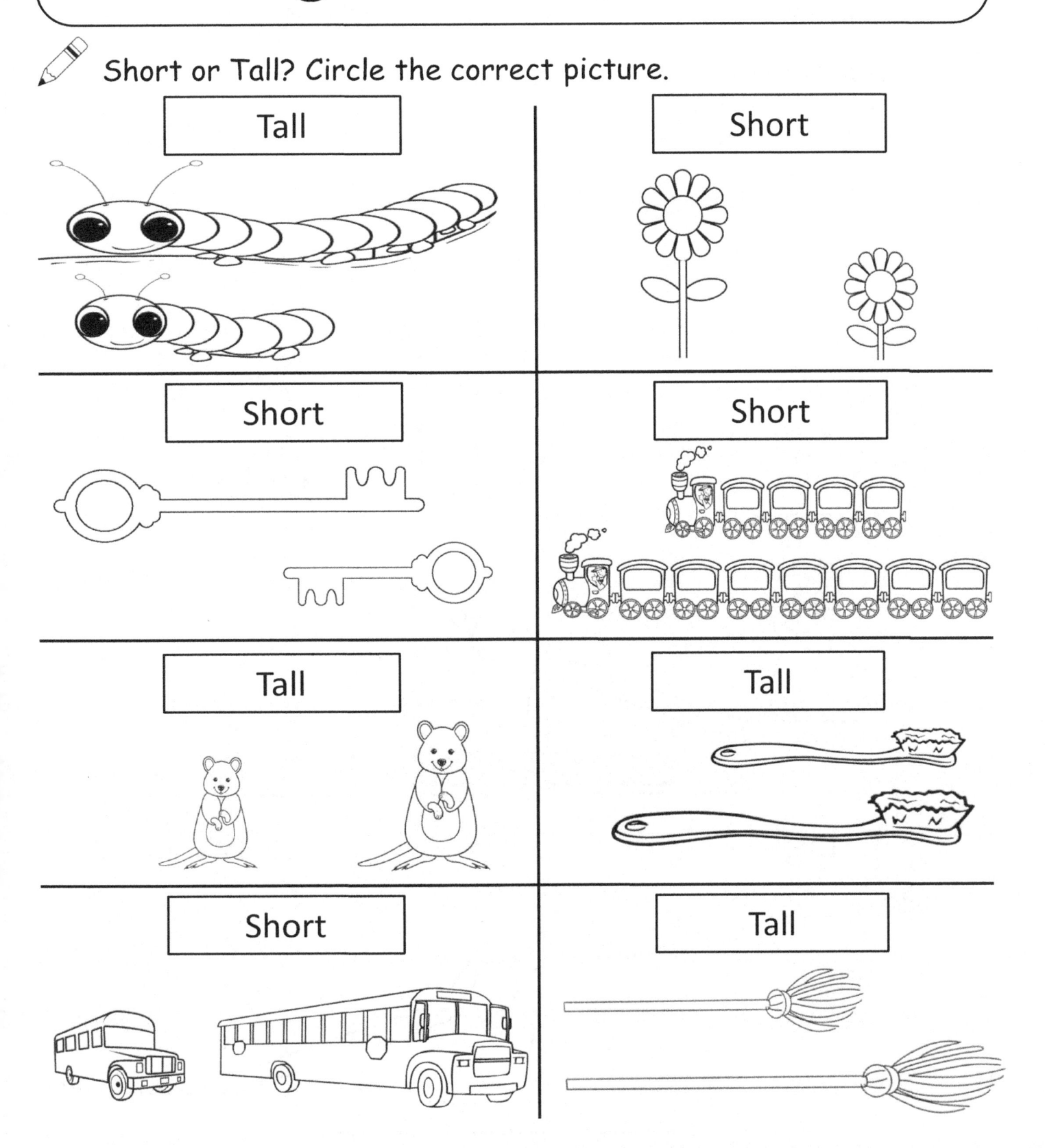

Short or Tall? Circle the correct picture.

Tall	Short
Short	Short
Tall	Tall
Short	Tall

Measuring
Weight : Heavy and Light

💡 Discover

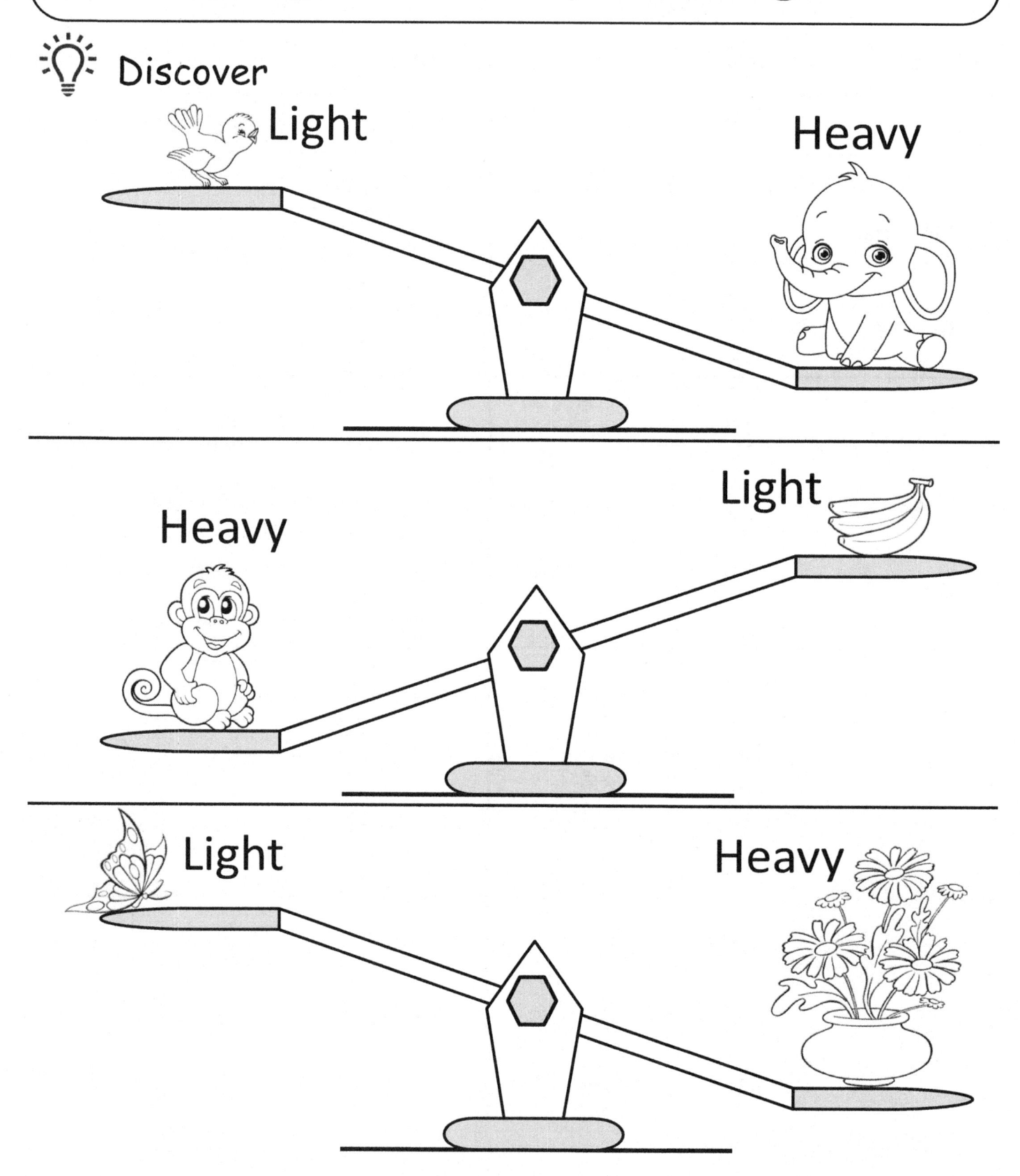

Light — Heavy

Heavy — Light

Light — Heavy

Measuring
Weight : Heavy and Light

✏️ Color the heavy one red and the light one yellow.

Measuring
Weight : Heavy and Light

✏️ Which object is heavy?

Measuring
Weight : Heavy and Light

✏️ Which object is light?

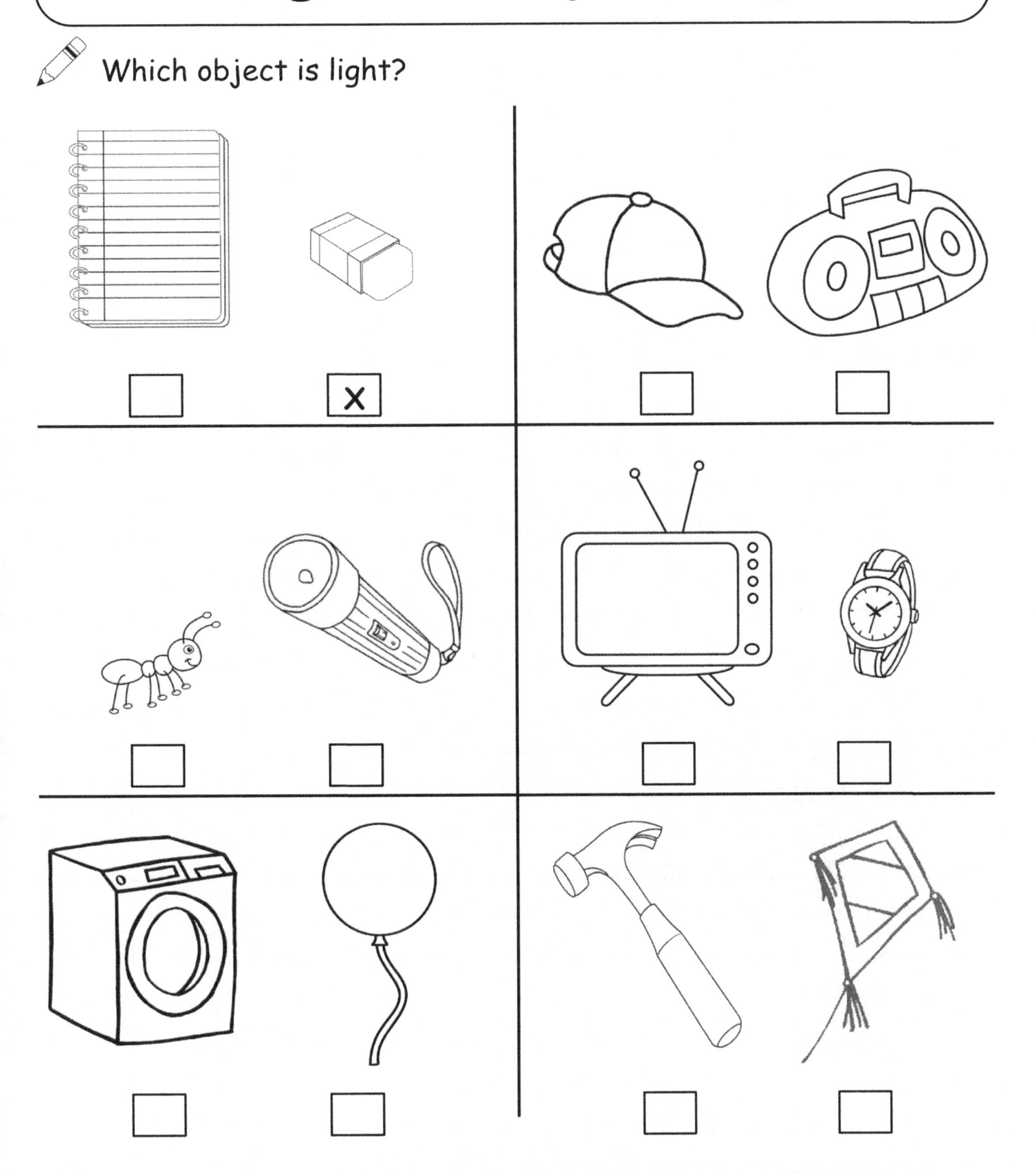

Measuring
Weight : Heavy and Light

Color the heavy one blue and the light one yellow.

Time

💡 Discover

Telling the time: We measure the time using Clocks

What time is it?

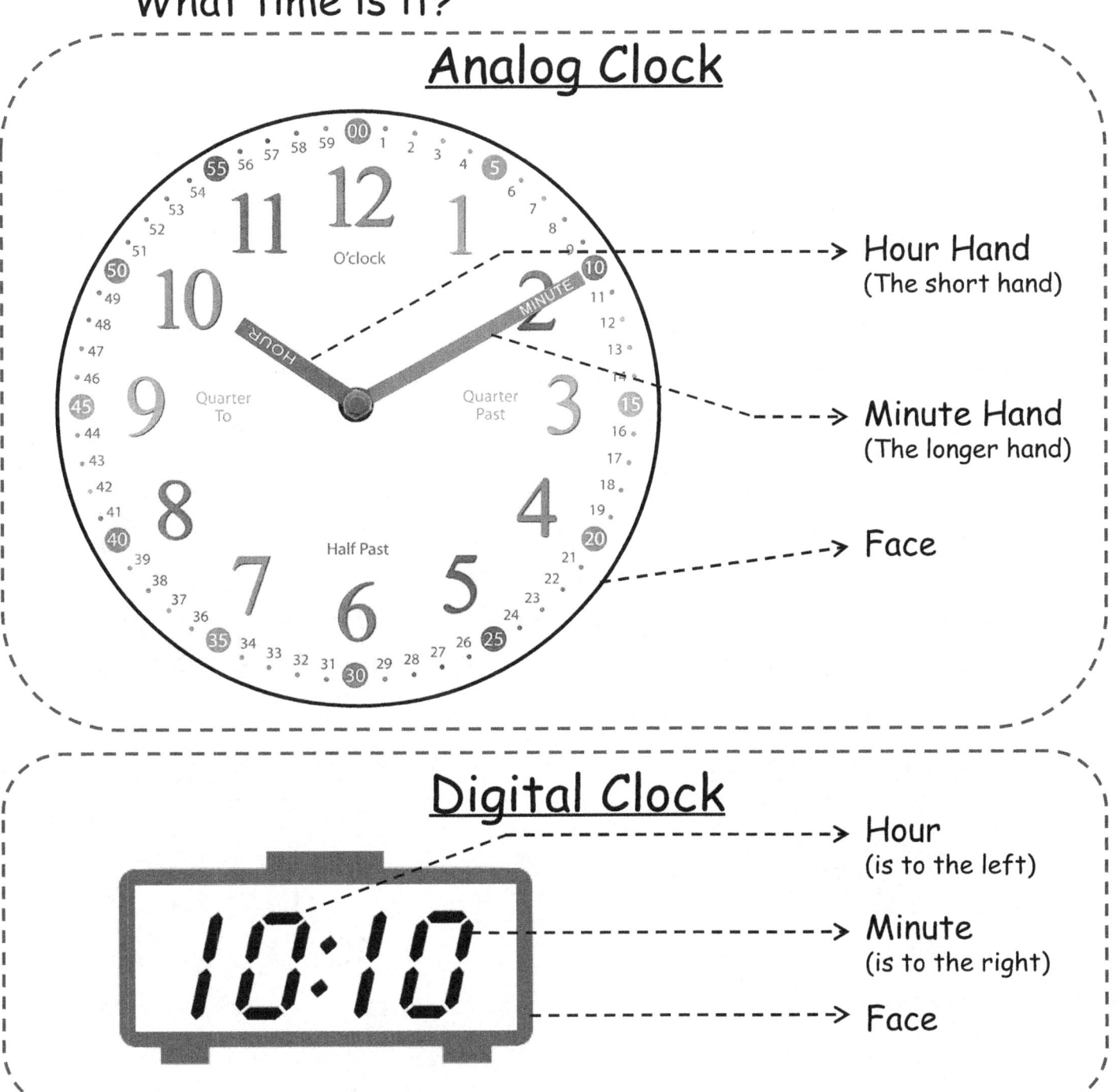

Analog Clock

- → Hour Hand (The short hand)
- → Minute Hand (The longer hand)
- → Face

Digital Clock

- → Hour (is to the left)
- → Minute (is to the right)
- → Face

What time is it? It's ten ten

There are 60 minutes in a hour

Time

💡 Discover

09:00 - It's nine o'clock

03:05 - It's three oh five

05:10 - It's five ten

06:30 - It's six thirty

03:45 - It's three forty-five

11:08 - It's eleven oh eight

Time

✏️ Match the times

1. [clock] • • 10:05 • • Five fifty-seven

2. [clock] • • 01:15 • • Ten oh five

3. [clock] • • 05:25 • • Eight ten

4. [clock] • • 08:10 • • Five twenty-five

5. [clock] • • 05:57 • • Nine o'clock

6. [clock] • • 09:00 • • One fifteen

Time

✏️ Write the correct time

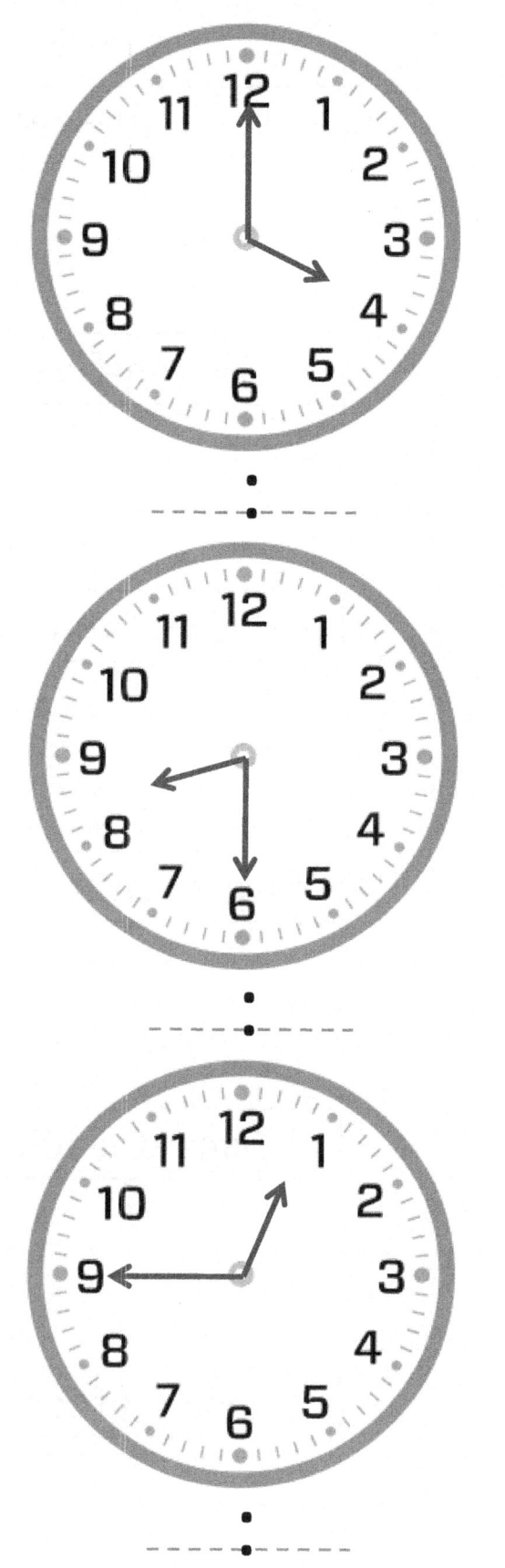

Time

Draw the hands on the clock to show the time

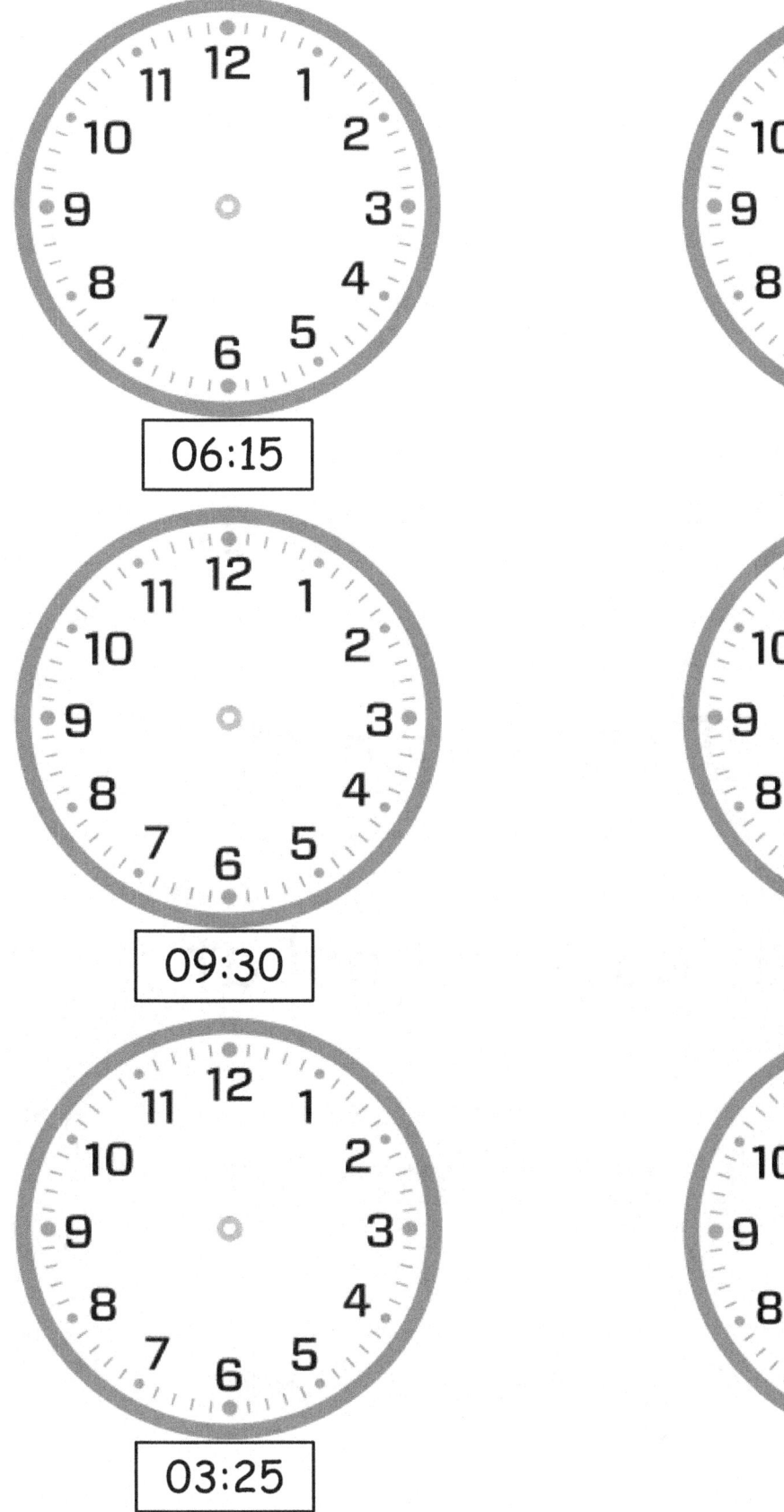

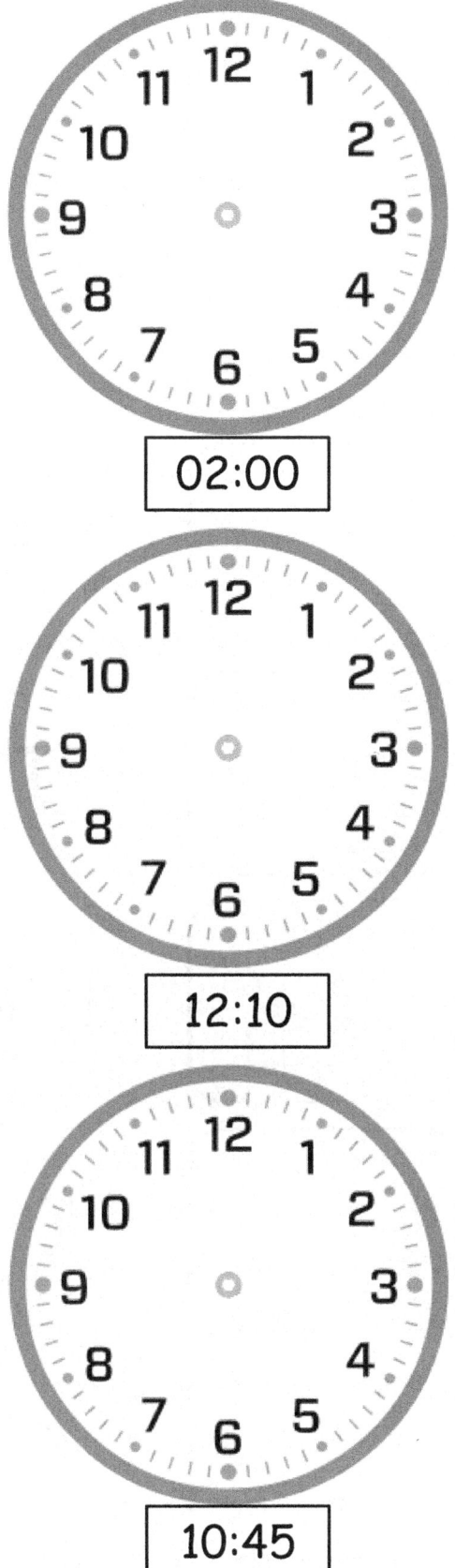

06:15

02:00

09:30

12:10

03:25

10:45

Time

My day...

1. I get up at

2. I eat breakfast at

3. I go to school at

4. I eat lunch at

5. I leave school at

6. I eat dinner at

7. I go to bed at

Hi!

Thank you for your trust, and we hope you will be completely satisfied with this kindergarten math workbook.

If so, and you have a few minutes, I would really appreciate you adding a comment to the Amazon site if you haven't already. It helps other parents make the right choice and encourages us to create more quality books.

As a small independent seller competing with much bigger companies, reviews help me stand out and help me see if I can improve my designs so I would appreciate it if you could take the time to leave one.
THANK YOU!

It is very easy to give me your opinion - just scan this QR Code. It will direct
you to the Amazon page where you can leave a review.

If for any reason whatsoever you are unhappy with your purchase, then please contact me **gloryworkpublishing@gmail.com** before leaving a review, so that I can resolve any problem that you might be facing.

Thanks once again, and have a lovely day ☺

Luciana

To discover our other creations, just scan this QR Code.
He will direct you to all of our books.

Do not hesitate to, "follow" us on Amazon so as not to miss any of the rest of our books.

All rights reserved © Glorywork Publishing

Made in the USA
Las Vegas, NV
10 May 2023